Mommsen and Cicero

FIRST EDITION NOT FOR SALE PRINTED AND
DISTRIBUTED WITH THE FINANCIAL CONTRIBUTION
OF THE NON-PROFIT ASSOCIATION <L'ITALIA FENICE>

"Theodor Mommsen" by Loescher & Petsch

Mommsen and Cicero

Vindiciae Ciceronianae

With a section on Ciceronianism,
Newtonianism and Eighteenth-Century Cosmology

by Vincenzo Merolle

Logos Verlag Berlin

Bibliographic information published by
the Deutsche Nationalbibliothek

The Deutsche Nationalbibliothek lists this publication in the
Deutsche Nationalbibliografie; detailed bibliographic data are
available on the Internet at http://dnb.d-nb.de .

Cover: Portrait of Cicero, marble sculpture, 1st century AD,
Capitoline Museum, Rome

ISBN 978-3-8325-3945-0

Logos Verlag Berlin GmbH
Comeniushof, Gubener Str. 47,
D- 10243 Berlin
Germany
Tel.: +49 (0)30 42 85 10 90
Fax: +49 (0)30 42 85 10 92
INTERNET: http://www.logos-verlag.com

To the memory of Giovanni Forni, eminent classical historian, my teacher at the high school in Arpino, who helped me find my own way at university.

Preface

Much has been written, and much has been said, about Cicero. A defence of the celebrated Roman statesman, orator, philosopher, against Theodor Mommsen's strictures, could even be considered superfluous, now that German historiography has recanted his views.

Nevertheless, we believe, something must be added on the subject of political philosophy and the influence of Ciceronian ideas particularly in the eighteenth century.

In the matter of political philosophy, the orator, philosopher, statesman, has substantially obscured the political writer in historical literature. The concept of a mixed constitution is in fact no more than modern liberalism. And, in contemplating the fall of the Roman Republic, we are led to think of the consequences of the defeat of the Girondins in France or of Kerensky in Russia. Neither of these two events contributed to the advancement of civilization.

In the matter of eighteenth-century cosmology, we have to observe a singular lack of knowledge of the classical world on the part of the scholars who work on the age of Enlightenment. We have done our best to fill the gap.

It remains to thank the friends and colleagues who have saved the present work from superficialities and errors. They are, in alphabetical order: Roger Emerson (London, Ontario), Michael Fry (Edinburgh), Maurizio Giovagnoli (Rome, 'La Sapienza'), Gianluca Gregori (Rome, 'La Sapienza'), Eugene Heath (New York), F.L. van Holthoon (Groningen), Vincent Hope (Edinburgh), Peter Jones (Edinburgh), Jürgen von Ungern-Sternberg (Basel), Andrea Luzzi (Rome, 'La Sapienza'), Piergiorgio Parroni (Rome, 'La Sapienza'), Robert Peevey (Arpino), Maria Romana Picuti (Foligno), Stefan Rebenich (Bern), Carlo Scappaticci (Arpino), Helmuth Schaaf (Ludwigshafen Rhein), Michael Sczekalla (Greifswald), Alessandra Tanzilli (Sora) and, last not least, Paolo Omodeo (Rome) who, with a generous grant of his non-profit association 'L'Italia Fenice', supported the whole Ciceronian project and rendered possible the printing of this book.

V. M.
Rome, December 2014

"Here one finds oneself in one of these places in history, that we observe with the same interest with which, in nature, we observe the sources and springs of the rivers, from which motion and life come through the countries. The science of Cicero can really be compared with the most majestic of the rivers of ancient literature... an immortal fame";[1]

"Arpino can therefore be called the Cyclop's cave of Roman imperial history";[2]

"In the same way as in the physical order of nature there are higher places from which we can command a view of the whole landscape, so other places command the panorama of history. Arpino is one of the most elevated of these places."[3]

---

[1] "Man befindet sich hier auf einer jener Stellen in der Geschichte, die man mit demselben Anteil aufsucht, wie in der Natur das steinerne Quellenhaus von Strömen, von denen Bewegung und Leben durch Länder und Zeiten kommt. Das Wissen Ciceros hat sich als ein Hauptstrom der alten Literatur ... ein unsterblicher Ruhm ...", Ferdinand Gregorovius, *Wanderjahre in Italien*, zweite Auflage (Leipzig, 1870), Zweiter Band, 'Lateinische Sommer. Von den Ufern des Liris', 1859, S. 259.

[2] "So ist Arpinum die wahre cyclopische Drachenhöhle der römischen Kaisergeschichte zu nennen", ibid., S. 260.

[3] "Wie gewisse Höhepunkte eine landschaftliche Aussicht dem Blick darbieten, so haben andere ein historisches Panorama um sich her. Arpinum ist ein solcher Höhepunkt", ibid., Ss. 260-61.

Contents

Contents

Introduction

This essay represents the answer of the city of Arpino to the unfair critical judgment given of the figure and literary work of Cicero by the celebrated German historian, Theodor Mommsen.

The author sketches a short history of the German anti-Ciceronianism, from Drumann to Mommsen and, after summarizing Mommsen's judgment, criticizes and counteracts it, observing that Cicero was faithful to the ideals of the Republic to the last moments of his life.

Sketching a short history of the 'liberal' literature, he demonstrates that the author of the *Römische Geschichte* was philosophically far from being a liberal, but was in fact a supporter of Caesarism. Therefore, he was unable to understand Cicero's concept of a mixed constitution, 'concordia ordinum' and republican ideals.

Nevertheless, during the last quarter of a century German historians have largely moved beyond Mommsen, producing at least seven biographies of Cicero[4], which ignore Mommsen altogether.

In the second section of the essay the author demonstrates that Newton's cosmology and eighteenth-century cosmology -see, in particular Adam Smith's philosophical works, and Hume's *The Natural History of Religion and Dialogues concerning Natural Religion*- are in fact nothing more than Cicero's cosmology, as discussed especially in *On the Nature of the Gods*. Therefore, Ciceronian philosophy was superseded -if it ever was- only with those of Kant and Hegel and, in part, with the physics of Maxwell and Einstein.

In appendix A, along with some other short writings, an 'Essai sur Cicéron' by the young Montesquieu, first published in 1892, and an essay by Moses S. Slaughter, severely critical of Mommsen, are included. The present author fully agrees with both Montesquieu and Slaughter, and considers it superfluous on his part to present a long defense of Cicero, since his points coincide with those of the two eminent writers.

In Appendix B the letters, almost 'early answers', to Michele Messina, are reprinted, written by the most eminent Italian and French historians of the late nineteenth century, all of them severely critical of Mommsen.

---

[4]See below, n. 24.

Appendix C is a short note on Cicero's birthplace with a bibliographical notice by Alessandra Tanzilli.

Last, the author expresses his wishes that a statue should be erected to Cicero, the defender of the Republic, in Via dei Fori Imperiali, in Rome, in front of the Curia.

PS: Latin texts of Cicero's works with English translations are those of the Loeb Classical Library.

Chapter I

Section I: Mommsen and Cicero

The sequel of insults that Theodor Mommsen uttered against Cicero in his *Römische Geschichte*, and that were soon answered by Gaston Boissier, could be defined as a matter of folklore. Boissier, in his *Cicéron et ses amis*, wrote that, to judge Cicero, to understand the real meaning of many of his pages, one needs "more acquaintance with life than one usually gets in German universities".[1]

So, why revisit this diatribe, one that subsequently involved the world of learning, and that, in 1873, directly concerned, as was inevitable, even this city, this Arpino of ours?[2]

First of all, let us start by saying that Mommsen was a great historian, a towering figure, the master of modern historiography on the Roman Republic, who deeply excavated in the real sources, which are epigraphy and archaeology, the narrations of ancient historians being just the general framework of history as it really was.

His *Corpus Inscriptionum Latinarum* remains as an imperishable monument to learning no less than his other works. His *Römische Geschichte*, which appeared in three vols in 1854-56, was the

---

[1] "Il faut avoir plus d'habitude de la vie qu'on ne prend d'ordinaire dans une université d'Allemagne", G. Boissier, *Cicéron et ses amis. Étude sur la société romaine du temps de César*, Hachette (Paris, 1865), repr. G. Olms Verlag (Hildesheim, 1976), p. 25.

[2] See Benedetto Croce, 'Il giudizio del Mommsen su Cicerone', in *Varietà di Storia Letteraria e Civile*, serie seconda, Laterza (Bari, 1949), pp. 1-12. In 1873 the city of Arpino printed at her own expense a pamphlet with the address by Abbot Mirabelli, then prof. of Latin literature in Naples, to Theodor Mommsen, kindly asking the celebrated German historian to reconsider his own judgment. The title of the pamphlet was *Theodorus Momsenius et M. Tullius Cicero. Prolusio habita in Archigymnasio Neapolitano XV Kal. Junias*, Neapoli, Ex Typographeo Unionis, 1873. The pamphlet has been reprinted with the title *Vindiciae Ciceronianae* (Arpino, 1984), c/o Vincenzo Zarrelli, with Latin text and Italian translation.

product of a superior mind, of a vigorous writer.[3] Nevertheless, his sharp and negative judgment of Cicero's figure needs to be examined more closely, more thoroughly, we believe, than it has ever been done in the past.

There is room for such a reappraisal, because the rich historiography of more recent times has paved the way for it, while a sort of embarrassment remains, on the part of the German biographers and apologists for Mommsen, at his unfair judgment on Cicero.

Over twenty centuries, about everything has been said on Cicero as a person and as a writer; it would be pointless to repeat much of that here. Therefore, appendix A reprints some particularly meaningful papers with which the present author fully agrees, that of the young Montesquieu in particular, along with a curious, bizarre letter, from Abbé Galiani to M.me D'Épinay.

In the appendix to the second edition of his *Apologia di Cicerone contro Teodoro Mommsen*,[4] Michele Messina published letters from a number of scholars, to whom he had sent a copy of the book, and who answered, giving various judgments of Mommsen as a historian and as a man. Their letters are a series of 'early answers' to his

---

[3]But let me make, here, a malicious remark, by transcribing the harsh judgment of Julius Beloch, who wrote that, in examining an inscription, Mommsen had had recourse to "reine Willkür oder vielmehr geradezu Fälschung", "pure arbitrariness or rather open falsification". See Beloch, *Römische Geschichte bis zum Beginn der punischen Kriege*, De Gruyter (Berlin und Leipzig, 1926), p. 515.

Julius Beloch (1854-1929) was professor of Roman History at the university of Rome. Mommsen long prevented him from having a chair in Germany, and only in 1912 he was appointed at the university of Leipzig. Ulrich von Wilamowitz-Moellendorff sided with him in the quarrels, sometimes violent, with Mommsen.

Johan Nicolai Madvig wrote on his part that "Mommsens römisches Staatsrecht ... trotz sehr vieler verdienstlicher Einzelheiten befriedigt mich doch im ganzen nicht", i.e. "Mommsen's public law ... despite many commendable particulars, in the whole does not satisfy me", J. N. Madvig, *Die Verfassung und Verwaltung des Römischen Staates*, Teubner (Leipzig, 1881), Vorwort, S. VIII.

[4]See second edition, (Napoli, 1882). The book by Messina, who writes well and well defends his theses, -for this reason its author was praised by Benedetto Croce- is virtually unobtainable, and from the internet only three Italian university libraries appear as in the possession of a copy. I am using the copy in the possession of the University of Cagliari, Sardinia.

Römische Geschichte and to his disparaging judgment of Cicero, and are here reprinted in the appendix B.

Last, but not least: an answer will be given to the question: was Mommsen a *liberal*, as he is usually defined by his German biographers, although they have never exhaustively explained this concept? Or was he, rather, a German nationalist? His 'liberalism', supposing that it is such, needs the test of a few concepts in the field of the history of political ideas, to understand its true nature.[5]

Drumann's Geschichte Roms

Mommsen's negative judgment had been to an extent prepared for by Wilhelm Drumann (1786-1844) in his *Geschichte Roms*.[6] Just before Drumann, Georg Barthold Niebuhr (1776-1831) had written that "Cicero followed truth in every way, and in his doing so we recognize the discord of his mind; he was in contradiction with himself". He then added: "I love Cicero as if I had known him and I judge of him as I would of a near relation who had committed a folly".

Moses Stephen Slaughter, commenting on the words of Niebuhr, wrote in *The Classical Journal*:

> "Cicero's 'folly' grew to a crime in the minds of Dru-
> mann, his most diabolical detractor, and of Mommsen,
> Drumann's unscrupulous successor in the business of
> character assassination. The crime for which Cicero is

---

[5] Although I am not a classical scholar, since this answer to Mommsen has long been overdue on the part of the city of Arpino, it is now my task to give it. I am in fact an Arpinas and in my boyhood I was a pupil at the 'Liceo Classico Tulliano', or classical high school, in Arpino. Mommsen's judgment haunted me since then.

[6] Wilhelm Drumann, *Geschichte Roms in seinem Übergange von der republikanischen zur monarchischen Verfassung oder Pompeius, Caesar, Cicero und ihre Zeitgenossen nach Geschlechtern und mit genealogischen Tabellen*, in six vols. First published in Königsberg in 1834-44, the work was later edited by P. Groebe, (Berlin-Leipzig, 1899-1929); see repr. G. Olms (Hildesheim, 1964). See in particular vol. V, pp. 218-697, and vol. VI, pp. 1-692, *Marcus Tullius Cicero*.

maligned by these anything but self-effacing critics is an incurable faith in a free state".[7]

This is the point at issue, as we shall demonstrate.

Drumann's work was certainly the "classic example of a party treatment" and "the most bizarre product of the German culture", as Eduard Meyer defined it.[8] Reading through the pages of the six big volumes, one comes to the conclusion that such an excessively detailed work can only with difficulty be called a work of history. It misses, in fact, the few general concepts which should lead the narration, concepts which are necessary to make a true historical work. Nevertheless, to its author's credit, it must be admitted that he points out a number of problems which have been developed and amplified in later historiography. He certainly deserves praise for this and for his diligent attitude, although almost 1400 pages on Cicero are too many indeed, and inevitably consist of no more than a paraphrase of Cicero's letters.

In general we can observe that one cannot do history just through the biographies of a few eminent characters.[9] As for Cicero, Drumann's 'character assassination' reaches its acme in the final part of vol. VI. If a primacy can be recognized for its author, it is that of having written, no doubt, the masterpiece, until now unequalled,[10] in this kind of literature. He never distinguishes among sources according to how defamatory they are of Cicero's character. He writes, for example, in mock seriousness:

---

[7]M. S. Slaughter, 'Cicero and his Critics', *Classical Journal*, 17.3 (1921-22), pp. 120-131, ibid., p. 123. Italics mine.

[8]"Musterbeispiel einer parteiischen Behandlung" and "das bizarrste Produkt deutscher Gelehrsamkeit", E. Meyer, *Caesars Monarchie*, Vorwort, S. VII.

[9]This is also the opinion of Mommsen who, referring to Drumann, criticizes the fact that "einige Historiker den verfehlten Plan nicht bloß gefaßt, sondern selbst teilweise ausgeführt haben, die Geschichte biographisch zu behandeln und ihre Hauptmomente nur an Personen anzuknüpfen", "some historians not only apprehended an inappropriate plan, but sometimes performed it in such a way as to deal with it from a biographical point of view and as to tie it only to persons", see Lothar Wickert, *Theodor Mommsen. Eine Biographie*, in vier Bänden, Klostermann (Frankfurt am Main, 1959-1980), Band I, S. 115.

[10]Only scurrilous can be defined J. Carcopino, *Les secrets de la Correspondence de Cicéron*, 2 vols (Paris, 1947). On this book see more here, in the *Bibliographical Note*.

"For the rest we know in his character to be excitability, selfishness, cowardice, defects of attention to justice and truth".[11]

The problem is how we take the numerous sentences like this in the narrative, since they hardly amount to a historical judgment; comment would surely be pointless. At any rate, it is hard to conceive a more malignant attitude in an author.

Drumann was motivated by conservative and monarchic principles, that he never disowned, and his own historiography was subservient to the demonstration of his theses. In the *Vorrede zum Gesamtwerk*, he wrote the following words:

"Outside my own intention my book is a work in praise of Monarchy, and I am pleased with this result, one I had not looked for, and such that it imposes itself not merely on Roman History, because a Prussian, a subject of Friedrich Wilhelm, cannot have any other creed than <Monarchy is the best>".[12]

This attitude explains in part his failings as a historian. His first volumes brought the reaction of the *London Quarterly Review*,[13] but nowadays even German scholars ignore the work as an embarrassment.

Drumann's unscrupulous successor

And "a problem for German cultural history"[14] are the words of Drumann's 'unscrupulous successor', Theodor Mommsen, who, unlike Drumann, is nevertheless, as said above, a brilliant writer. In

---

[11] Vol. VI, p. 369, & 419: "Übrigens erkennt man in seinem Charakter Erregbarkeit, Selbstsucht, Feigheit und Mangeldefect an Achtung vor Recht und Wahrheit"

[12] "Nicht wider, aber ohne meinen Willen ist mein Buch ein Lobschrift auf die Monarchie, und ich freue mich des nicht gesuchten Ergebnisses, welches sich mir nicht bloß in der Römischen Geschichte aufdringt, denn der Preusse, der Untertan eines Friedrich Wilhelm, kann kein anderes politisches Glaubensbekenntnis haben als ἡ μοναρχία κράτιστον, die Monarchie ist das stärkste".

[13] The article is reprinted here in appendix A.

[14] "Ein Problem der deutscher Kulturgeschichte", Walter Rüegg, *Cicero und der Humanismus. Formale Untersuchung über Petrarca und Erasmus* (Zürich, 1946), *Vorwort*, S. VIII.

his pages, the reader's attention is attracted in particular by the mastery of the style, by the full command of the subject with which its author deals. The words of Anthony Grafton, in his Introduction to an English edition of the *History of Rome*, appear appropriate:

> "The flaws and contradictions in his work, most of them now easy to detect, were and are a small price to pay for its fresh, comprehensive and vivid portrait of a civilization taking shape".[15]

As a consequence of Mommsen's pages "nowadays Cicero in Germany is almost unknown or rejected as a representative of a liberal humanity", as Walter Rüegg complained in 1946.[16] And, let us add, the problem is so embarrassing that Mommsen's biographers have preferred to ignore it -so Hartmann, Heuß and Wickert-, or just to touch on it incidentally -it being impossible to ignore it- like the most recent Stefan Rebenich.[17]

Slaughter concluded his essay of 1922, cited above, with the following words:

> "Have we not the right to demand of the German historians to set a less biased man to the task of re-writing the story of the Roman Republic, one who knows neither ira nor studio, and may we not expect at his hands a

---

[15] Abridged edition, The Folio Society (London, 2006), p. XXI. Also Nietzsche denounced Mommsen's method: "The author who seeks to make Roman history come alive by loathsome references to the paltry views of modern political parties and their ephemeral configuration creates a greater sin against the past than does the mere scholar who leaves everything dead and mummified", cited by Grafton, ibid. And Gaston Boissier, in his *Cicéron et ses amis*, op. cit., explained that "il est assez d'usage aujourdh'ui d'aller demander à l'histoire du passé des armes pour la lutte du présent ... il ne faut jamais oublier que c'est d'outrager l'histoire que de la mettre au service des intérêts changeants des partis, et qu'elle doit être, suivant la belle expression de Thucydide, une œuvre faite pour l'éternité", Boissier, ibid., p. 28.

[16] "Heute ist Cicero in Deutschland so gut wie unbekannt oder dann als Vertreter eines liberalen Meschentums verworfen", Rüegg, op. cit., ibid., S. VIII.

[17] Ludo Moritz Hartmann, *Theodor Mommsen. Eine biographische Skizze*, Perthes (Gotha, 1908); Lothar Wickert, see above, note 14; Alfred Heuß, *Theodor Mommsen und das 19. Jahrhundert*, Franz Steiner Verlag (Stuttgart, 1996; Nachdruck der Ausgabe Hirt, Kiel, 1956); Stefan Rebenich, *Theodor Mommsen. Eine Biographie*, Verlag C. H. Beck (München, 2007).

fairer treatment of the man whose unpardonable sin was a belief in free institutions?".[18]

After the embarrassed admission of Walter Rüegg, it has taken a long time, for the German historians, to admit that Mommsen's influence prejudiced a fair treatment of the figure of Cicero, of his role and influence in the history of the late Roman Republic, and in the world of learning.

The process of revision, almost forty years ago, marked an essential step forward with Manfred Fuhrmann.[19] He tried to give an explanation of why Mommsen's words were so harsh against 'Cicero der Advokat', but the explanation, almost a justification, was quite embarrassed, and so substantially dissatisfying. According to Fuhrmann, Mommsen's words are essentially owed to the tradition of *Rhetorikverachtung*, or contempt of rhetoric, typical of the German tradition, of which Mommsen was also somehow 'a prisoner'. Its influence he was unable to escape.

Fuhrmann did not mention Drumann at all and did not touch on the subject of the political ideas of Mommsen, therefore his explanation was not exhaustive, and only in part convincing. As a scholar he is well known for his works on rhetoric[20] and for his translation of Cicero's works into German, but the history of social and political ideas was distant from his interests. Nevertheless he concluded his paper first of all distinguishing the true rhetoric from the false one, and saying that Cicero himself "has obviously had a feeling, that true rhetoric is possible only in a state and within given institutions".[21] Therefore, a "new attitude towards rhetoric", that the German world of learning had finally discovered, was likely to have created the premises, also for Germany, for discovering Cicero the orator. With these words Fuhrmann was referring to his *Cicero und die Römische Republik. Eine Biographie*, that was then

---

[18]Slaughter, op. cit., p. 130.

[19]M. Fuhrmann, 'Die Tradition der *Rhetorik*-Verachtung und das deutsche Bild vom <Advokaten> Cicero', in *Rhetorik*, Max Niemeyer Verlag (Tübingen, 1989), pp. 43-55.

[20]See his book *Die Antike Rhetorik*, 4. Aufl. (Zürich, 1995).

[21]"Cicero selber hat offensichtlich ein Empfinden dafür gehabt, daß wahre Rhetorik nur in einem Staatswesen und innerhalb gegebenen Institutionen möglich sei", 'Die Tradition', p. 55.

about to appear.[22] The book actually concluded by complaining that Germany "recently has been the scene of an unhappy Cicero reception", …"while this negative attitude, effectively propagated by Mommsen, had the result that the significance of Cicero seemed to exhaust itself in formalism".[23]

After Fuhrmann's book, German scholars have abundantly compensated, the most recent historiography being a substantial recantation,[24] although the shadow of Mommsen still hovers as a ghost in the air, and the only way to deal with the matter remains that of ignoring him, avoiding to mention him at all. This is, obviously, the way chosen by most recent historians.

Mommsen's judgment of Cicero's figure and work

Let us go on here by reprinting Mommsen's judgment on Cicero, in order to try to understand the reasons for it. Mommsen's judgment is written, no doubt, in a vigorous prose, that shows how vigorous

---

[22] "Es wäre denkbar, daß die neue Ausstellung zur Rhetorik die voraussetzungen dafür geschaffen hat, daß auch Deutschland den Redner Cicero entdecke", ibid. The book first appeared in Mannheim in 1989.

[23] "Für deutsche Leser soll allerdings noch angefügt werden, daß ihr Land unlängst Schauplatz einer recht unglücklichen Cicero-Rezeption gewesen ist", while "diese negative, besonders effektvoll von Mommsen propagierte Betrachtungsweise bewirkte, daß sich die Bedeutung Ciceros im formalem zu erschöpfen schien", ibid., p. 310.

[24] See in particular Matthias Gelzer, *Cicero. Ein Biographischer Versuch* (zweite Auflage, Wiesbaden, 1983); Ch. Habicht, *Cicero der Politiker* (Münich, 1990); Manfred Fuhrmann, *Cicero und die Römische Republik. Eine Biographie*, Artemis & Winkler (fifth ed., 2011); Jürgen Leonhardt, *Ciceros Kritik der philosophischen Schulen* (München, 1999), Zetemata Heft 103; Stephanie Kurczyk, *Cicero und die Inszenierung der eigenen Vergangenheit*, Böhlau (Köln, 2006); Klaus Bringmann, *Cicero* (Darmstadt, 2010); Wilfried Stroh, *Cicero: Redner, Staatsmann, Philosoph*, Beck (München, 2008, second ed. 2010); Arnd Morkel, *Marcus Tullius Cicero*, Konigshausen & Neumann (Würzburg, 2012); Wolfgang Schuller, *Cicero oder der letzte Kampf um die Republik. Eine Biographie*, Beck (München, 2013). Furthermore *Marco Tulio Cicerón*, by Francisco Pina Polo, Ariel Publishing (Barcelona, 2005), has been published in a German edition by Klett-Cotta Verlag with the title *Rom. Das bin ich. Marcus Tullius Cicero. Ein Leben*, and the subtitle *Die Cicero-Biographie für das 21. Jahrhundert. Aus dem Spanischen übersetzt von Sabine Panzram* (Stuttgart, 2010). Therefore, we can say that it has been 'adopted' by the German world of learning.

was the mind of the celebrated German historian, who was certainly "kein borussischer monarchist wie Drumann".[25]

A) "On the advocate's platform -the only field of legal opposition left open by Sulla ... the adroit speaker Marcus Tullius Cicero (born 3rd January 106), son of a landholder of Arpinum, speedily made himself a name by the mingled caution and boldness of his opposition to the dictator".[26]

B) "The leader of the compliant majority continued to be Marcus Cicero. He was useful on account of his lawyer's talent of finding reasons, or at any rate words, for everything; and there was a genuine Caesarian irony in employing the man ... as the mouthpiece of servility".[27]

"Even Cicero, however humbly he always bowed before the regents, issued an equally invenomed and insipid pamphlet against Caesar's father-in-law".[28]

C) "Marcus Tullius Cicero (106-43) who was from the outset quite as much author as forensic orator; he published his pleadings regularly, even when they were not at all or but remotely connected with politics. This was a token not of progress, but of an unnatural and degenerate state of things. Even in Athens the appearance of non-political pleadings among the forms of literature was a sign of debility; and it was doubly so in Rome, which did not like Athens ... the exaggerated pursuit of rhetoric, but borrowed it from abroad arbitrarily and in antagonism to the better traditions of the nation ... he was the creator of the modern Latin Prose; his importance rests on his mastery of style, and it is only as a

---

[25] K. Bringmann, op. cit., p. 16.
[26] I am citing from the English edition *The History of Rome*, translation by W. P. Dickson, (London 1894, repr. Bristol 1996), ibid., book V, chap. VIII, p. 118.
[27] Ibid., book V, chap. VIII, p. 132.
[28] Ibid., book V, chap. VIII, p. 135.

stylist that he shows confidence in himself. In the character of an author, on the other hand, he stands quite as low as in that of a statesman. He essayed the most various tasks, sang the great deeds of Marius and his own petty achievements in endless exameters, beat Demosthenes off the field with his speeches, and Plato with his philosophic dialogues; and time alone was wanting for him to vanquish also Thucydides. He was in fact so thoroughly a dabbler, that it was pretty much a matter of indifference to what work he applied his hand. By nature a journalist in the worst sense of that term - abounding, as he himself says, in words, poor beyond all conception in ideas -there was no department in which he could not with the help of a few books have rapidly got up by translation or compilation of a readable essay. His correspondence mirrors most faithfully his character. People are in the habit of calling it interesting and clever, and it is so, as long as it reflects the urban or villa life of the world of quality; but where the writer is thrown on his own resources, as in exile, in Cilicia, and after the battle of Pharsalus, it is stale and empty as was ever the soul of a feuilletonist banished from his familiar circles. It is scarcely needful to add that such a statesman as such a littérateur could not, as a man, exhibit aught else than a thinly varnished superficiality and heartlessness. Must we still describe the orator? The great author is also a great man ... Cicero had no conviction and no passion, he was nothing but an advocate, and not a good one. He understood how to set forth his narrative of the case with piquancy and anecdote, to excite, if not the feeling, at any rate the sentimentality of his hearers, and to enliven the dry business of legal pleading by cleverness or witticisms mostly of a personal sort; his better orations, though they are far from coming up to the free gracefulness and the sure point of the most excellent compositions of this sort, for instance the *Memoirs* of Beaumarchais, yet form easy and agreeable reading. But while the very advantages just indicated will appear to the serious judge as advantages of very dubious value; the absolute want of political discernment in the orations or constitutional questions and of juristic deduction in the forensic addresses, the egotism forgetful of its duty and constantly losing sight of the cause while thinking of the advocate, the dreadful barrenness of thought in the Ciceronian orations must revolt every reader of feeling and judgment.

If there is anything wonderful in the case, it is in truth not the orations, but the admiration that they excited. As to Cicero every unbiased person will soon make up his mind: Ciceronianism is a problem, which in fact cannot be properly solved, but can only be resolved into the great mystery of human nature -language and the effect of language on the mind ... The Ciceronian manner ruled no doubt throughout a generation the Roman advocate-world ... but the most considerable men, such as Caesar, kept themselves always aloof from it ... They found Cicero's language deficient in precision and chasteness, his jests deficient in liveliness, his arrangement deficient in clearness and articulate division, and above all his whole eloquence wanting in the fire which makes the orator."[29]

D) As for the *De Oratore*, *De Republica* and *De Legibus*, Mommsen continues saying that "they are no great works of art, but undoubtedly they are the works in which the excellence of the author are most, and his defects least, conspicuous. The treatise *De Republica* carries out, in a singular mongrel compound of history and philosophy, the leading idea that the existing constitution of Rome is substantially the ideal state-organization sought for by the philosophers, an idea indeed just as unphilosophical as unhistorical ... The scientific groundwork of these rhetorical and political writings of Cicero belongs of course entirely to the Greeks, and many of the

---

[29] Ibid., book V, chap. XII, pp. 503-507.

details also, such as the grand concluding effect in the treatise *De Republica*, the Dream of Scipio, are directly borrowed from them; yet they possess comparative originality, inasmuch as the elaboration shows throughout Roman colouring, and the proud consciousness of political life, which the Roman was certainly entitled to feel as compared with the Greeks, makes the author even confront his Greek instructors with a certain independence. The form of Cicero's dialogue is doubtless neither the genuine interrogative dialectics of the best Greek artificial dialogue nor the genuine conversational tone of Diderot or Lessing ...

While these rhetorical and political writings of Cicero with a philosophic colouring are not devoid of merit, the compiler on the other hand completely failed, when in the involuntary leisure of the last years of his life (45-44) he applied himself to philosophy proper, and with equal peevishness and precipitation composed in a couple of months a philosophical library. The receipt was very simple. In rude imitation of the popular writings of Aristotle,[30] in which the form of dialogue was employed chiefly for the setting forth and criticizing of the different older systems, Cicero stitched together the Epicurean, Stoic, and Syncretist writings handling the same problem ... and all that he did on his own part was, to supply an introduction prefixed to the new book from the ample collection of prefaces for future works which he had beside him ... In this way no doubt a multitude of thick tomes might very quickly come into existence -'They are

---

[30] In the German edition of the *Römische Geschichte* (see Deutscher Taschenbuch Verlag, München, 1976), fünftes Buch, 12 Kapitel, III, 623, also occurs the word 'aristotelisch'. This can appear rather surprising, because Cicero's source of inspiration is in Plato's *Dialogues*, only in part in Aristotle. Nevertheless Cicero writes, in a letter to Atticus, "itaque cogitabam, quoniam in singulis libris utor prohemiis ut Aristoteles in iis quos ἐξωτερικούς vocat, aliquid efficere ut non sine causa istum appellarem", "so I am thinking of making a suitable occasion to address him (i.e., Varro) in one of the prefaces which I am writing to each book, as Aristotle did in what he calls his 'exoteric' pieces", Cicero to Atticus, letter 89, 2. Aristotle's dialogues, although known in Cicero's times, haven't been transmitted to posterity.

copies', wrote the author himself to a friend who wondered at his fertility; 'they give me little trouble, for I supply only the words and these I have in abundance'.

Against this nothing further could be said; but anyone who seeks classical productions in works so written can only be advised to study in literary matters a becoming silence".[31]

Cicero, a noble soul

Obviously, we will not 'study' the 'becoming silence' demanded by Mommsen, willingly admitting that his portrait of Cicero is a literary masterpiece, but a bold one, in which its author proves how vigorous his mind was, how powerful the force of his imagination.[32] Nevertheless, it has very little to do with history, and needs to be considered just as a brilliant display of its author's talent as a writer.

To refute his unfair and bizarre judgment of Cicero's person and work, we don't need to write a eulogy, or repeat what has been said, better than we can, by other, distinguished historians.[33] On Cicero as a writer, an answer to the words of Mommsen is out of the question. As for his place in the history of the Roman Republic and of political ideas, we just need to listen directly to the orator's words, to realize how great his true personality was.

Let us start with the *Second Philippic*, with the *divina Philippica*, as Juvenal[34] singled it out.

> "In these twenty years there was never an enemy of the
> Republic who did not at the same time declare war on

---

[31] *History of Rome*, pp. 508-510.

[32] I willingly admit, with Karl Christ, that "die Gediegenheit der Kenntnisse, die Sichereit des Urteils, Passion und suggestiver Stil des Autors haben bewirkt, daß diese Darstellung nichts von ihrem Glanz, ihrem künstlerischen und wissenschaftlichen Rang verloren hat. Die *Römische Geschichte* ist und bleibt ein Höhepunkt unserer Geschichtschreibung": see K. Christ, 'Theodor Mommsen und die *Römische Geschichte*', in Mommsen, *Römische Geschichte*, (München, 1976), Band 8, S. 7. Nevertheless I am not a classical scholar, and my objections are from the point of view of the history of ideas, of social and political ideas in particular.

[33] See here, App. A.

[34] Juvenal, *Tenth Satire*, line 125.

me too... none of those people became my enemy by choice; they were all challenged by me for the sake of the Republic";[35]

"I defended the Republic when I was young; I shall not desert it now that I am old ... and I should be happy to offer my body if my death can bring into reality the freedom of our state";[36]

As for Caesar's person:

"Deluded wretch, with never in his life a glimpse of even the shadow of Good";[37]

"how can he behave otherwise than as an infamous man? That is precluded by his life, his character, his past, the nature of his present enterprise, his associates, the strength of the honest men or just their persistence";[38]

As for Caesar's death:

"for never, holy Jupiter, was a greater deed done in Rome or anywhere else in the world, none more glorious, none more sure to live forever in the memory of mankind";[39]

"Caesar had talent, the ability to reason, a retentive memory, literary talent, concentration, reflection, industry. His military achievements, even though disastrous

---

[35] "Nemo his viginti rei publicae fuerit hostis qui non bellum eodem tempore mihi quoque indixerit ... Nemo enim illorum inimicus mihi fuit voluntarius, omnes a me rei publicae causa lacessiti", *Phil.* II, i, 1.

[36] "Defendi rem publicam adulescens, non deseram senex ... Quin etiam corpus libenter obtulerim, si repraesentari morte mea libertas civitatis potest.", *Phil.* II, xxxvi, 118-19.

[37] "Oh hominem amentem et miserum, qui ne umbram quidem umquam τοῦ καλοῦ viderit!", Ad Atticum, 21 January 49, Loeb Classical Library, letter no. 134.

[38] "Qui hic potest se gerere non perdite? Vetat vita, mores, ante facta, ratio suscepti negoti, socii, vires bonorum aut etiam constantia", Ad Atticum, 8 March 49, Loeb Classical Library, letter no. 169.

[39] "Quae enim res umquam, pro sancte Iuppiter, non modo in hac urbe sed in omnibus terris est gesta maior, quae gloriosior, quae commendatior hominum memoriae sempiternae?", *Phil.* II, xiii, 32.

to the Republic, had been great. Aiming at monarchy for many years ... he had cajoled the naive populace with shows, with buildings, with gifts and with feasts ... his adversaries by a show of clemency ... he had already succeeded in habituating a free community to servitude ... brave men ... have learned what a beautiful thing it is to slay a tyrant ... what fame and glory it brings ... from now on it will be a race to get to the job".[40]

And from the *De Officiis*, where the general treatment is philosophical, but the passion of the writer erupts in harsh judgments:

"there is no social relation ... all more close, none more dear than that which links each one of us with our country. Parents are dear; dear are children, relatives, friends; but one native land embraces all our loves; and who that is true would hesitate to give his life for her, if by his death he could render her a service? So much the more execrable are these monsters who have torn their fatherland to pieces with every form of outrage and who are and have been engaged in compassing her utter destruction";[41]

"... when everything passed under the absolute control of a despot ... when I had lost the friends who had been associated with me in the task of serving the interests of the state, and who were men of the highest standing,

---

[40] "Fuit in illo ingenium, ratio, memoria litterae, cura, cogitatio, diligentia; res bello gesserat, quamvis rei publicae calamitosas, at tamen magnas; multos annos regnare meditatus, magno labore, magnis periculis quod cogitarat effecerat; muneribus, monumentis, congiariis, epulis multitudinem imperitam delenierat; suos praemiis, adversarios clementiae specie devinxerat.... quam sit re pulchrum, beneficio gratum, fama gloriosum tyrannum occidere ... Certatim posthac, mihi crede, ad hoc opus curretur neque occasionis tarditas exspectabitur", *Phil.* II, xxxxv-xxxxvi, 116-118.

[41] "Sed ... omnium societatum nulla est gravior, nulla carior quam ea, quae cum re publica est uni cuique nostrum. Cari sunt parentes, cari liberi, propinqui, familiares, sed omnes omnium caritates una complexa est, pro qua quis bonus dubitet mortem oppetere, si ei sit profuturus? Quo est detestabilior istorum immanitas, qui lacerarunt omni scelere patriam et in ea funditus delenda occupati et sunt et fuerunt", *De Officiis*, I, xvii, 57. Cicero has in mind Caesar, Clodius, Catiline.

29

I did not resign myself to grief, by which I should have been overwhelmed, had I not struggled against it, neither, on the other hand, did I surrender myself to a life of sensual pleasure unbecoming to a philosopher ... government ... had fallen into the hands of men who desired not so much to reform as to abolish the constitution".[42]

With concepts and a wording that terribly apply to the experience of the men of our century:

"... *no amount of power can withstand the hatred of the many.* The death of this tyrant, whose yoke the state endured under the constraint of armed force, and whom it still obeys more humbly than ever, though he is dead, illustrates the deadly effects of popular hatred, and the same lesson is taught by the similar fate of *all other despots, of whom practically no one has escaped such a death* ... Those who in a free state deliberately put themselves in a position to be feared are the maddest of the mad ... *Freedom suppressed and again regained bites with keener fangs than freedom never endangered*";[43]

"... the present victor ... so great was his passion for wrong-doing that the very doing of wrong was a joy to him for its own sake, even when there was no motive for it".[44]

---

[42] "... cum autem dominatu unius omnia tenerentur ... socios denique tuendae rei publicae, summos viros, amisissem, nec me angoribus dedidi, quibus essem confectus, nisi iis restitissem, nec rursum indignis homine docto voluptatibus. Atque utinam res publica stetisset, quo coeperat, statu nec in homines non tam commutandarum quam evertendarum rerum cupidos incidisset!", *De Officiis*, II, i, 2-3.

[43] "Multorum autem odiis nullas opes posse obsistere ... Nec vero huius tyranni solum, quem armis oppressa pertulit civitas ac paret cum maxime mortuo, interitus declarat, quantum odium hominum valeat ad pestem, sed reliquorum similes exitus tyrannorum, quod haud fere quisquam talem interitum effugit... qui vero in libera civitate ita se instruunt, ut metuantur, iis nihil potest esse dementius ... Acriores autem morsus sunt intermissae libertatis quam retentae", *De Officiis*, II, vii, 23-24. Italics mine.

[44] "At vero hic nunc victor ... tanta in eo peccandi libido fuit, ut hoc ipsum eum delectaret, peccare, etiamsi causa non esset", *De Officiis*, II, xxiv, 84.

As for the killing of the tyrant, with words and concepts quite similar to these that in the Middle Ages will recur in St. Thomas, who considers the tyrant as a wild beast, that must be opposed and killed:

> "No creature more vile or horrible than a tyrant, or more hateful to gods and men, can be imagined; for, though he bears a human form, yet he surpasses the most monstruous of the wild beasts in the cruelty of his nature";[45]

> "The Greeks accord divine honours to those men who have slain despots. What sights have I seen at Athens and in other cities of Greece. What religious rites ordained in their honour!";[46]

> "... if anyone kills a tyrant ... The Roman People ... of all glorious deeds they hold such an one to be the most noble";[47]

> "... we have no ties of fellowship with a tyrant, but rather the bitterest feud ... a man whom it is morally right to kill; -nay, all the pestilent and abominable race should be exterminated from human society";[48]

> "Behold, here you have a man who was ambitious to be king of the Roman People and master of the whole world; and he achieved it! The man who maintains that such an ambition is morally right is a madman; for he

---

[45] "tyrannus, quo neque taetrius neque foedius nec dis hominibusque invisius animal ullum cogitari potest; qui quamquam figura est hominis, morum tamen immanitate vastissimas vincit beluas...", *De Republica*, II, xxvi, 48.

[46] "Graeci homines deorum honores tribuunt eis viris, qui tyrannos necaverunt: quae ego vidi Athenis! Quae aliis in urbibus Graeciae! Quas res divinas talibus institutas viris!", *Pro Milone*, xxix, 80.

[47] "... si quid tyrannum occidit ... Populo quidem Romano ... ex omnibus praeclaris factis illud pulcherrimum existimat", *De Officiis*, III, iv, 19.

[48] "Nulla est nobis societas cum tyrannis, et potius summa distractio est ... quem est honestum necare, atque hoc omne genus pestiferum atque impium ex hominum communitate exterminandum est ... sic ista in figura hominis feritas et immanitas beluae ...", *De Officiis*, III, vi, 32.

justifies the destruction of law and liberty and thinks
their hideous and detestable suppression glorious ... For
ye immortal gods! can the most horrible and hideous
of all murders -that of fatherland- bring advantage to
anybody, even though he who has committed such a
crime receives from his enslaved fellow-citizens the title
of 'father of his country'?".[49]

As for the despotism of the multitude, that recalls to the mind
of the reader the 'tyranny of the majority', dealt with by Alexis de
Tocqueville in his *Democracy in America*:

" ... despotism of the multitude ... such a gathering ...
is just as surely a tyrant as if it were a single person, and
an even more cruel tyrant, because there can be nothing
more horrible than that monster which falsely assumes
the name and appearance of a people".[50]

We confess that it is extremely demanding to comment on the
words of Cicero, of this noble soul, of this 'hater of tyrants'. Some
of his expressions should be simply sculpted in marble.[51] One needs
to conclude that in historical literature *his are the noblest and most
luminous words one has ever read on the subject of liberty*. We
challenge here the historians of social and political ideas to find
something similar in the great authors of the past. We will find
sometimes similar words and concepts, often taken from him, – in
the works of the great Montesquieu, for example- but certainly never
expressed with his vigour and passion. Only such a noble soul as his
own could pronounce and write these words, and it is not clear to

---

[49] "Ecce tibi, qui rex populi Romani dominusque omnium gentium esse concu-
piverit idque perfecit! probat enim legum et libertatis interitum earumque
oppressionem taetram et detestabilem gloriosam putat. Potest enim, di im-
mortales! cuiquam esse utile foedissimum et taeterrimum parricidium patriae, quamvis is, qui se eo obstrinxerit, ab oppressis civibus parens nomine-
tur?", *De Officiis*, III, xxi, 83. The title of pater patriae was bestowed not
only on Cicero for saving the Republic against Catiline, in 63, but also on
Caesar after the battle of Munda, in 45.

[50] "... in multitudinis dominatu ... sed est tam tyrannus iste conventus, quam si
esset unus, hoc etiam taetrior, quia nihil ista, quae populi speciem et nomen
imitatur, immanius belua est", *De Rep.*, III, xxxiii, 45.

[51] See for example here above, lines emphasized in text corr. with note 43.

us how it is possible to defame a man who feels such noble ideals so deeply, ideals for which he suffered martyrdom. Such a man should be simply admired, even sanctified, because he was one of the few men, very few men, indeed, born to greatness.

With the experience of twenty centuries of history behind us, with the recent memories of Hitler and Stalin, whose crimes led to the destruction of entire populations and upset the natural course of human society, we could well imagine what life was like in Rome, in one of the most turbulent periods of history, when the beheading of opponents was the daily practice, or the rule, regularly applied by the winners.

Therefore, it would be a waste of time, even unflattering, to insist on refuting Mommsen, a bizarre man, a giant in philology but a dwarf in politics. And it would be a waste of time to answer the other detractors of Cicero, whose soul was too great to be vulnerable to defamation on the part of little men, unable to conceive noble ideals.[52]

If for Christians he is 'the pagan Christian' -'not a Pagan philosopher but a Christian Apostle', as Petrarch defined him-, for the believers in the ideals of liberty he is the 'noble father', to be raised to the glory of altars.

Far from being a 'tremulous hero',[53] this was Cicero.

Nevertheless we need to go on, in the demonstration of our thesis.

Mommsen's personal character

Before going on with the demonstration of the theses of this essay, we should briefly consider Mommsen's personal character, his im-

---

[52] We refer, in particular, to the hack writers, who, after his death, prostituted their talent in defaming Cicero, and run to the service of the winners. It is unflattering for Mommsen the fact that he could write such a bizarre judgment of Cicero's person and work, without reflecting on the nobility of Cicero's great soul.

[53] As from *This was Cicero. Modern Politics in a Roman Toga*, Secker & Warburg, (London, & Norwood, Mass., 1942), by H. J. Haskell, and *The Tremulous Hero. The Age and Life of Cicero. Orator, Advocate, Thinker and Statesman*, Pallas Publishing (London, 1939), by A. F. Witley. Both books, although the work of non-professional historians, deserve praise for their literary merits.

petuous nature which was intolerant of anybody and of anything that contradicted him.

See, for example, his letter, dated Breslau, 9 Sept. 1856, to Wilhelm Henzen, then first secretary of the Deutsches Archäologisches Institut in Rome, where he writes the following, contemptuous words:

> "But as we cannot help Cicero, so even less can we help those who complain for him, for as a rule they are complaining under the scourge of their own churlishness. Just leave them and regard them as *praeficae*, who follow a funeral".[54]

Soon after arriving to Rome, on 30 Dec. 1844, his manner gave offence in the *Istituto di Corrispondenza Archeologica*. On the occasion of the publication of some inscriptions by two influential scholars, Raffaele Garrucci and Gianpietro Secchi, both Jesuits, he set off an embarrassing quarrel, that caused a sensation in Germany as well, and it cost the *Istituto* some diplomatic effort to soothe the resentments that he had aroused. On this occasion he showed himself "uneinsichtig", lacking any insight, showing contempt for his opponents, calling them *Pfaffenvolk* and *Lausezeug*, 'priest-ridden people' and 'lice'.[55]

In the Roman salon of Countess Caetani Lovatelli he saw Ferdinand Gregorovius, the celebrated author of the *Geschichte der Stadt Rom im Mittelalter* (1859-1872), and addressed him: "Can I give you some advice? Write a History of Rome in the Middle Ages".

The point is that Gregorovius' *Geschichte* had been published some years before. Although received with a certain coolness in the German academic world, the work is a literary masterpiece, and is based on original research. It is written with a splendid literary

---

[54] "Aber wie dem Cicero nicht zu helfen ist, so ist es noch weniger denen, die um ihn jammern, denn sie jammern in der Regel um ihre eigene unter dieser Geißel mitleidende Miserabilität. Lassen Sie sie nur und betrachten (to look out) Sie sie als die *praeficae*, die dem Leichenbegängnis folgen", Wickert, Band III, pp. 641-2, *Anmerkungen*.

[55] See Rebenich, op. cit., p. 49. See also C. Ferone, "Raffaele Garrucci nelle corrispondenze di Th. Mommsen, F. Ritschl, E. Gerhard", <Rend. Acc. Napoli> LXII (1989-90), pp. 33 ff.

approach, peculiar to its author and, from this point of view, it is incomparable.

In his *Diario Romano*, in 1862, Gregorovius wrote the following words: "Mommsen's work is remarkable for scholarship, for critical and destructive acumen, but it is more a libel than a history".

He added, in 1873: "Mommsen ... like Richard Wagner, is ill with megalomania. The <professors of the chair> don't want to recognize me, because I work in independence, I don't hold any bureaucratic appointment, and, *horribile dictu*, I possess some poetical talent".[56]

During his triumphal visit to Sardinia, in a dinner organized in his honour in Cagliari, on 16 Oct. 1877, Mommsen gave out incautious, unfair, contemptuous judgments, which caused the local newspaper, the *Avvenire di Sardegna*, to speak of the 'germano', 'invidioso tedesco', 'incalzante orda germanica'.[57] In the next, triumphal dinner in Sassari, on 26 October, he was nevertheless less polemical than usual, apparently because on that occasion he was not drunk. In any case, he did not omit to speak of 'camorra, da me combattuta', "racket, mafia style, that I have fought against".[58]

When Ulrich von Wilamowitz-Moellendorff, who had been his student in Berlin, and had married his daughter Marie, submitted to him a translation of Euripide's *Hyppolitos*, he rejected it with objections expressed in his habitual, harsh words, that "kränkten Wilamowitz sehr", that hurt Wilamowitz very much, who afterwards always complained of Mommsen's 'herrische Caesarnatur'.[59]

---

[56] "Posso darvi un consiglio? Scrivete una storia di Roma nel Medio Evo", see Antonio Muñoz, 'Ferdinando Gregorovius e le sue opere', introd. to *Storia della città di Roma nel Medioevo* (Città di Castello, 1938), vol. I, pp. XVII-XVIII.

[57] The adjective 'germano' has in Italian literary tradition a polemical meaning, even if not necessarily a derogatory one. See Manzoni, in *Marzo 1821*, published in 1848, strophe 9, lines 6-8: "quel Dio/ che non disse al Germano giammai: /va', raccogli ove arato non hai/ spiega l'ugne, l'Italia ti do". The poem, published in 1848, exhorting the Italians to the war of independence against Austria, renders the true mood of the Italians against the Germans, particularly against Austria, that occupied Northern Italy, Milan and Venice in particular, long preventing the unification of the Italian nation.

[58] On the journey to Sardinia see A. Mastino, 'Il viaggio di Teodoro Mommsen e dei suoi collaboratori in Sardegna per il '*Corpus Inscriptionum Latinarum*", in *Theodor Mommsen e L'Italia*, Accademia Nazionale dei Lincei (Roma, 2004), pp. 225-344, passim.

[59] Rebenich, op. cit., p. 207.

Apart from this, there were, obviously, differences in political convictions. 'Die politischen Manifeste seines Schwiegervaters', the political manifestos of his father-in-law were 'unerträglich', unbearable to Wilamowitz, while Mommsen made it clear, in a letter to Lujo Brentano, that his son-in-law was 'keineswegs mit mir gleicher politischer Gesinnung', in no way of similar political outlook to him.

Another of his victims was Julius Beloch,[60] who was long disappointed in his hope of getting a chair in a German University, because of Mommsen's opposition.

Mommsen's 'masterpiece', from the point of view of his intemperance of character, was shown when, in 1881, he was elected for the Linksliberalen in the National Parliament. He attacked the all-powerful Chancellor Bismarck in the Reichstag, accusing him of political fraud. In consequence he had to face legal action, but was discharged by a well-disposed tribunal, with the finding that his were political, not personal allegations.

Finally, in 1897, in a letter to the *Neue Freie Presse* of Vienna, he called the Czechs "apostles of barbarism", adding that "the Czech skull is impervious to reason, but susceptible to blows".

Mommsen, a 'liberal' historian?

The peculiarities of Mommsen's strong character explain, although only in part, the vigour of his polemical account of Cicero's person and work. Nevertheless, considering him as a 'liberal' tout court, as his German apologists take for granted, is inadmissible. His support of Napoleon III and Caesar's authoritarian democracy is in fact an unresolved contradiction.

He was certainly 'a man of the 1848', which was the year of the 'liberal' revolutions in Europe, the time when nationalities fought against the 'old regime', and he took part in 'liberal' movements with the spirit of all the generous young people of his own time. Nevertheless the problem is whether he was a real liberal, or only a German subject fighting for 'liberal' struggles during his early years. And in any case, also in the case of Mommsen, we have to assume that there should be what we call 'a perfect correspondence between general, philosophical ideas and political commitment'. In

---

[60]On Beloch see above, n. 3.

sum, Mommsen's personality scarcely coincides with that of a 'liberal'. After all, one can take part in a 'liberal' movement for a particular aim, without necessarily being a real 'liberal', in a general, philosophical sense.

There is, in reality, an essential difference between political and philosophical liberalism, the first one being merely contingent, i.e., depending on particular, historical circumstances. For example, one can well oppose a dictatorship, but if the aim is that of replacing it with another one, one is far from being a liberal. Philosophical liberalism, by contrast, is a conception of life and history, which goes beyond a particular event.

Mommsen's biographers[61] have had recourse to the Hegelian concept of 'historical necessity', which justified -historically, not morally, it must be clear- the dictatorship of Caesar, but this concept is both abstract and philosophical, not political, and we are not sure that Mommsen had so elaborated his own ideas from a philosophical point of view.

The *Römische Geschichte* moves from the fundamental idea that Caesar was the demigod who saved the Roman state from breakdown. All his opponents were consequently miserable people, who vainly tried to oppose the new demiurge.

Mommsen was the supporter of what we call authoritarian democracy, enlightened despotism, *democratic despotism, Caesarism* or, in a quite recent neologism, *presidentialism*, meaning, with this word, a particular form of the political institutions in which a President, once he is elected directly by the people, is superior to the legislature or parliament, which cannot check or restrain his deeds. Obviously, the next step is from 'presidentialism' to autocracy, and then to despotism and dictatorship.

---

[61]See in particular Heuß, op. cit., pp. 131 ff., who makes any possible effort to find Hegelianism in Mommsen, forgetting, nevertheless, that the 19th century was the century of the 'nations', and that liberalism is much more than the German concept of 'Nation'. He writes as follows: "Mommsen hat sicherlich liberal und demokratisch gedacht ... ist die Geschichte als ganzes die Entfaltung sittlicher Energien und Darstellung der Freiheit, unter welchen Aspekten auch immer sie sich zeigt ..." (*"Mommsen certainly used to think liberal and democratic ... history as a whole is the development of spiritual energies and the representation of the liberty, and it shows herself under these aspects"*). This is a Hegelian concept, no doubt, but not a Mommsenian one, we fear.

But despotism is never 'enlightened', as we have learned, and every day learn even more from history and politics, and *'authoritarian democracy'* is never 'enlightened'. The force of democracy, its vitality, as Alexis de Tocqueville explained in his *Démocratie en Amérique*, is in fact slow to move, but becomes an irresistible force when penetrating the conscience of a society, when a society has finally realized the relevance of events and the necessity of taking part in them.

Cicero as a political writer. His main concepts

The history of European liberalism is a series of elaborations by the great writers, who wrote in connection with particular historical events. So John Locke's *Two Treatises of Government*, that are the philosophical and political justification of the 'glorious revolution' of 1688. Montesquieu, the master of eighteenth-century liberalism, wrote under and against the despotism of the *ancien régime*. So Constant, who opposed Napoleon's despotism in the name of the 'romantic' ideas of liberty. Alexis de Tocqueville was the most advanced of the theorists of democratic liberties and, among other relevant authors, -including, in particular, John Stuart Mill- Benedetto Croce, the 'philosopher of liberty'.

The Aristotelian and Ciceronian concept of a mixed constitution as being the best, as composed of monarchic, popular and aristocratic elements, paradoxically is the most modern, the most advanced of all, in the sense that the ideal constitution is representative of all the social strata or classes, not to say individuals, not one being excluded. In this way it becomes *the true palestra of liberty*, to which, although with its own contradictions, nothing can be superior. It allows, in fact, historical change without any hindrance, in a most natural, and therefore in a lasting way.[62]

---

[62]Typically Thomas Hobbes, the author of the *Leviathan*, judges Aristotle and Cicero with the following words: "It is an easy thing, for men to be deceived, by the specious name of Libertie ...when the same errour is confirmed by the authority of men in reputation for their writings in this subject, it is no wonder if it produce sedition, and change of Government. In these westerne parts of the world, we are made to receive our opinions ... from Aristotle, Cicero ... that living under Popular States, derived those Rights, not from the Principles of Nature ... so Cicero, and other writers, have grounded their Civill doctrine, on the opinions of the Romans, who were taught to hate

Nevertheless, Cicero's contribution from the point of view of political thought has been neglected by historians, with the exception of a couple of books, the work of talented historians, no doubt, but limited in their extent and aim, and in their perception of what politics is.[63]

It can be considered a paradox on our part to consider Cicero, as we are doing in the present section of our essay, as the most perfect 'liberal' author of the history of political thought. And, certainly, the word 'liberal' appears in the English vocabulary only in the fourteenth century. Nevertheless the texts speak for themselves. Therefore, let us hear directly from Cicero.

As for the concept of liberty, with words that recall to the memory of the reader Dante's Cato 'libertà va cercando ch'è così cara/come sa chi per lei vita rifiuta':[64]

> "... liberty has no dwelling place in any State except that in which the people's power is the greatest, and surely nothing can be sweeter than liberty; *but if it is not the same for all*, it does not deserve the name of liberty".[65]

As for 'exaggerated licence' which produces tyranny, with words and concepts that will recur in Edmund Burke for the French revolution:

Monarchy ... And by reading these Greek, and Latine Authors, men from their childhood have gotten a habit (under a false shew of Liberty), of favouring tumults, and of licentious controlling the actions of their Soveraigns ... with the effusion of so much blood", *Leviathan*, Part II, chap. 21, p. 267 of the Penguin edition (Harmondsworth, 1981).

[63] See in particular Fr. Cauer, *Ciceros politisches Denken* (Berlin, 1903), and Neal Wood, *Cicero's social and political thought*, University of California Press (Berkeley and Los Angeles, 1988); but see also H. Strasburger, *Concordia Ordinum. Eine Untersuchung zur Politik Ciceros* (Borna, 1931, repr. Amsterdam 1956).

[64] "Of liberty he journeys: that how dear/ They know, who for her sake have life refused", Dante, *The Divine Comedy, Purgatorio*, I, 71-2. Translation by Henry Francis Cary.

[65] "... et talis est quaeque res publica, qualis eius aut natura aut voluntas, qui illam regit. Itaque nulla alia in civitate, nisi in qua populi potestas summa est, ullum domicilium libertas habet; qua quidem certe nihil potest esse dulcius, et quae, si aequa non est, ne libertas quidem est", *De Rep.*, I, xxxi, 47. Italics mine.

"from this exaggerated licence ... tyrants spring up as from a root ... so liberty itself reduces people who possess it in too great degree to servitude ... this extreme liberty gives birth to a tyrant and the utterly unjust and cruel servitude of tyranny. For out of such an ungoverned, or rather, untamed, populace someone is usually chosen as leader... some bold and depraved man ... To such a man, because he has much reason to be afraid if he remains a private citizen, official power is given and continually renewed ... If the better citizens overthrow such a tyrant ... then the state is re-established ... oligarchy is only a tyranny of another kind".[66]

As for superiority of mixed constitution:

"... kingship, in my opinion, is by far the best of the three primary forms, but a balanced form of government which is a combination of the three good simple forms is preferable even to kingship";[67]

"I consider the best constitution for a state to be that which is a balanced combination of the three forms mentioned, kingship, aristocracy and democracy...";[68]

"... as far as the ideal state is concerned ... I defined the three commendable types of States and the three bad types which are their opposites ... I demonstrated that no single one of three types is the ideal, but that

---

[66] "ex hac nimia licentia, quam illi solam libertatem putant, ait ille ut ex stirpe quadam existere et quasi nasci tyrannum ... itaque ex hac maxima libertate tyrannus gignitur et illa iniustissima et durissima servitus. Ex hoc enim populo indomito vel potius immani deligitur ... audax, inpurus, consectans proterve ... cui quia privato sunt oppositi timores, dantur imperia et ea continuantur ... quos si boni oppresserunt ... recreatur civitas; sin audaces, fit illa factio, genus aliud tyrannorum", *De Rep.*, I, xliv, 68.

[67] "Ex tribus primis generibus longe praestat mea sententia regium, regio autem ipsi praestabit id, quod erit aequatum et temperatum ex tribus optimis rerum publicarum modis", *De Rep.*, I, xlv, 69;

[68] "... statu esse optimo constitutam rem publicam, quae ex tribus generibus illis, regali et optumati et populari ...", *De Rep.*, II, xxiii, 41.

a form of government which is an equal mixture of the three good forms is superior to any of them by itself";[69]

"Our ancestors, when the rule of kings had become intolerable to them ... created magistracies to be held for a year only, with the restriction, that the Senate was set up as a Council over the State for ever, and they ordained that the members of that Council should be chosen by the whole people, and that industry and merit should open the way for admission to that exalted Order for all citizens".[70]

With words and concepts that will be adopted by Montesquieu he treated the function of the 'corps intermédiaires' in society:

"... riches, names and power, when they lack wisdom and the knowledge of how to live and to rule over others, are full of dishonour and insolent pride, nor is there any more depraved type of state than that in which the richest are accounted the best ... Thus, between the weakness of a single ruler and the rashness of the many, aristocracies have occupied that *intermediate position which represents the utmost moderation*";[71]

As for democracy or popular government and equality, with concepts and words that are fundamental in the theories of liberalism, theories which believe in the free competition both of individuals

---

[69] "de optimo autem statu equidem arbitrabar ... definieram genera civitatum tria probabilia, perniciosa autem tribus illis totidem contraria, nullumque ex eis unum esse optimum, sed id praestare singulis, quod e tribus primis esset modice temperatum", *De Rep.*, II, xxxix, 65.

[70] "... a maioribus nostris sapientissime constitutam, qui cum regum potestatem non tulissent, ita magistratus annuos creaverunt, ut consilium senatus rei publicae praeponerent sempiternum, diligerentur autem in id consilium ab universo populo aditusque in illum summum ordinem omnium civium industriae ac virtuti pateret", *Pro Sestio*, lxiv, 137.

[71] "Nam divitiae, nomen, opes vacuae consilio et vivendi atque aliis imperandi modo dedecoris plenae sunt et insolentis superbiae, nec ulla deformior species est civitatia quam illa, in qua opulentissimi optimi putantur ... Sic inter infirmitatem unius temeritatemque multorum medium optimates possederunt locum, quo nihilo potest esse moderatius", *De Rep.*, I, xxxix, 51-52. Italics mine.

and of ranks in society, and oppose the equality imposed with co-
ercive methods:

> "For equality of legal rights of which free people are
> so fond cannot be maintained ... and what is called
> equality is really most inequitable. For when equal hon-
> our is given to the highest and the lowest ... then this
> very 'fairness' is most unfair; but this cannot happen in
> States ruled by their best citizens";[72]

> "liberty ... does not consist in serving a just master, but
> in serving no master at all";[73]

Again with words and concepts that will occur almost word for
word in Montesquieu:

> "Of course he (the statesman) should be given almost
> no other duties than this one ... of improving and exam-
> ining himself continually, urging others to imitate him,
> and furnishing in himself, as it were, a mirror to his
> fellow-citizens by reason of the supreme excellence of
> his life and character".[74]

Last, with a statement in which the historians of political ideas
will recognize the fundamental concept of common law and eighteenth-century polemic about legislators, projectors, founders of

---

[72] "Nam aequabilitas quidem iuris, quamquam amplexantur liberi populi, neque
servari potest ... eaque, quae appellatur aequabilitas, iniquissima est", *De
Rep.*, I, xxxv, 53.

[73] "... libertas, quae non in eo est, ut iusto utamur domino, sed ut nullo", *De
Rep.*, II, xxiii, 43

[74] "Video iam, illum, quem expectabam, virum cui praeficias officio et muneri ...
ut numquam a se ipso instituendo contemplandoque discedat ...", *De Rep.*,
II, xlii, 69. See Montesquieu: "Le gouvernement de Rome fut admirable
en ce que ... tout abus du pouvoir y pût toujours être corrigé ... Le gouvernement d'Angleterre ... il y a un corps qui l'examine continuellement, et
qui s'examine continuellement lui-même; et telles sont ses erreurs qu'elles
ne sont jamais longues, et que, par l'esprit d'attention qu'elles donnent à la
nation, elles sont souvent utiles.
En un mot, un gouvernement libre, c'est-à-dire toujours agité, ne sauroit
se maintenir, s'il n'est, par ses propres lois, capable de correction", Montesquieu, *Considérations sur les causes de la Grandeur des Romains et de
leur Décadence*, chap. VIII, ed. La Pléiade, (Paris, 1951), p. 115-16.

states, occurring especially in the works of Adam Smith and of Adam Ferguson:[75]

> "Cato used to say that our constitution was superior to those of other States on account of the fact that almost every one of these other commonwealths had been established by one man, the author of their laws and institutions; for example, Minos in Crete, Lycurgus in Sparta ... On the other hand our own commonwealth was based upon the genius, not of one man, but of many; it was founded, not in one generation, but in a long period of several centuries and many ages of men. For, said he, there never has lived a man possessed of so great genius that nothing could escape him, nor could the combined powers of all the men living at one time possibly make all necessary provisions for the future without the aid of actual experience and the test of time".[76]

Modern theories of liberalism

In modern theories of liberalism, we can distinguish those that believe in the freedom of the individual and others that believe in the concept of the alternation of majority and minority in government, the minority in its own turn becoming a majority when it wins the consent of society, and vice versa.

Benjamin Constant, for example, was the head of the parliamentary opposition to the French governments of the Bourbon restoration, after Napoleon's fall, and his were 'romantic' ideas of liberty.

Nevertheless let me cite here Benedetto Croce who, in his *Manifesto degli Intellettuali Antifascisti* (*Il Mondo*, 1 May 1925), in op-

---

[75]For these concepts in Smith and Ferguson see here, below, text corr. with notes 81-91.

[76] "(Cato) dicere solebat ob hanc causam praestare nostrae civitatis statum ceteris civitatibus, quod in illis singuli fuissent fere, qui suam quisque rem publicam constituissent legibus atque institutis suis, ut Cretum Minos, Lacedaemoniorum Lycurgus...... nostra autem res publica non unius esset ingenio, sed multorum, nec una hominis vita, sed aliquot constituta saeculis et aetatibus. Nam neque ullum ingenium tantum extitisse dicebat, ut, quem res nulla fugeret, quisquam aliquando fuisset, neque cuncta ingenia conlata in unum tantum posse uno tempore providere, ut omnia complecterentur sine rerum usu ac vetustate", *De Rep.*, II, i,1.

position to the praise of 'unanimity' of Gentile's *Manifesto degli Intellettuali Fascisti*, (appearing in *Il Popolo d'Italia*, National Fascist Party, 21 April 1925), offered a specimen of his philosophy, which was essentially a development of Hegelianism. He wrote that

> "the essence of liberalism lies in an historical concept of free competition, including alternating parties in power, one consequence of which is that progress is realized through opposition and change."[77].

Croce's views, as the philosopher himself declared, were mainly "inspired and took shape from resistance against the oppression of fascism",[78] but its philosophical origins were in Hegel's dialectic method.

As for Tocqueville's democracy, or *liberalism of the minorities*, this theory is suspicious of the concept of *majorities*. In the analysis of American society the concept of the '*tyranny of the majority*' has a particular relevance. As the author writes, "si jamais la liberté se perd en Amérique, il faudra s'en prendre à l'omnipotence de la majorité qui aura porté les minorités au désespoir."[79]

The fundamental principle of Tocqueville's *liberalism of the minorities*, with particular reference to the United States of America, is that all minorities have their own liberties, not one that attempts to threaten the liberties of the others, because they know that their own liberty depends on the liberty of the others. Consequently, the real danger is that of the tyranny of the majority.

Furthermore:

> "Si la démocratie a plus de chances de se tromper qu'un roi ou un corps de nobles, elle a aussi plus de chances de revenir à la vérité, une fois que la lumière lui arrive, parce qu'il n'y a pas, en général, dans son sein, d'intérêts contraires à ceux du plus grand nombre"; "les lois de la démocratie tendent, en général, au bien du plus

---

[77]For quite similar concepts see B. Croce, 'La concezione liberale come concezione della vita', in *Etica e Politica*, Laterza (Bari, 1956), pp. 290-300.

[78]Croce, 'Di un libro sulla libertà in Italia', in *Scritti e Discorsi Politici*, II, pp. 311-5, ibid., p. 312.

[79]*Démocratie en Amérique*, Gallimard, I, vii, 'Tyrannie de la Majorité', pp. 261-4, but these words occur on p. 271.

grand nombre, car elles émanent de la majorité de tous les citoyens, laquelle peut se tromper, mais ne saurait avoir un intérêt contraire à elle-même"; "il y a plus de lumières et de sagesse dans beaucoup d'hommes réunis que dans un seul".[80]

Last, Montesquieu:

"Pour règle générale, toutes les fois qu'on verra tout le monde tranquille dans un État qui se donne le nom de République, on peut être assuré que la liberté n'y est pas ... Ce qu'on appelle union dans un corps politique, est une chose très équivoque ... Il peut y avoir de l'union dans un État où on ne croit voir que du trouble";[81] *"c'est une expérience éternelle que tout homme qui a du pouvoir est porté à en abuser ... pour qu'on ne puisse abuser du pouvoir, il faut que, par la disposition des choses, le pouvoir arrête le pouvoir".*[82]

This last sentence can be considered as the right answer to Mommsen's 'authoritarian democracy'.

Political and philosophical liberalism

As we have seen, there is a substantial distinction between political and philosophical liberalism, and we have to emphasize the point that *philosophical* liberalism incorporates *all the moments of the 'reality'*, all the elements of the contradiction, the so called category of totality, or *Gesamtheit*, using Hegelian vocabulary.

Political liberalism is to be considered no more than an ideology, when it represents *one of the moments of the contradiction, one of the moments composing the whole.* But *philosophical* liberalism moves from the concept that we cannot simply *cancel* a part of the 'reality' as an error. Only Liberalism, in a philosophical sense, contains *both the moments of the contradiction*, that of progress and that of conservation. Therefore, it constitutes a superior synthesis,

---

[80]Tocqueville, ibid. deuxième partie, V, p. 235, VI, p. 242, VII, 258.
[81]Montesquieu, *Considérations*, in *Œuvres Complètes*, II, II, p. 119.
[82]Montesquieu, *Esprit des Lois*, XII chap. iv. Italics mine.

because it represents *the consciousness of history* itself, and cannot be replaced by an inferior moment.

A mature consciousness of what liberalism is, as a synthesis of eighteenth and early nineteenth-centuries philosophical and political ideas, appeared only with Hegel, Constant, later with Tocqueville and other authors, when reflection about past history became mature, the concept of a synthesis was reached, and human mind could command, and theorize, centuries of history. Once more Hegel was right, when he says that philosophy is like the owl of Minerva, that flies out only at sunset: i.e., it is a reflection about past history, not on the present. As for present events, we are in fact just *actors on the stage of history.*

As for the nineteenth century and, in general, from a philosophical point of view, we should consider, in a superior synthesis, not only the category of *Enlightenment* but also that of *Romanticism,* which *opposed* and *criticized* Enlightenment, in this way making *progress*, and improving on it.[83]

The alternative is that of considering an author, or a historical period, in an 'abstract and philosophical light',[84] of theorizing abstract categories as eternal and immutable, coercing historical knowledge and superimposing abstract categories on history. This perspective leads to a new kind of 'Enlightenment', which is a meta-historical category, therefore absolute, not admitting of contradictions and rejecting the principle that a society can change *naturaliter*, for the worse, but also for the better.

The *truth*, in a philosophical sense, and particularly in politics, is in fact *problematic*,[85] and *cannot be achieved once and forever.*

---

[83]I am echoing concepts from Benedetto Croce's writings, as repeatedly in the present section of this essay.

[84]The wording is borrowed from Adam Smith, *Theory of Moral Sentiments* (Oxford UP, 1979), VII, III, chap. 1, p. 316. The *Theory* was first published in 1759.

[85]Paraphrasing words and concepts from Benedetto Croce, Antonio Gramsci, the theorist of Italian Marxism, wrote that "in the scientific debate, since we suppose that the interest is the pursuit of the verity and the progress of science, shows himself more *advanced* he who puts himself from the point of view that the adversary can express an exigency that must be incorporated, although as a subordinate moment, in his own construction.

Understanding and valuing realistically the positions and the reasons of the adversary ... means emancipating oneself from the prison of ideologies

And so it is in historical literature, for it to be actually *science*, and not *ideology*: i.e., if it wants to serve knowledge, and not simply points of view, or to ideology of particular social classes or ranks.

Therefore the general, philosophical idea, the concept of *totality*, of the *whole*, or *Gesamtheit*, should lead to this particular conclusion, avoiding any absurd pretence of replacing what cannot be replaced, as the whole cannot be replaced by a part.

The risk is that an *intellectual*, -from the Latin *intelligĕre*-, should be nothing more than what could be called 'a warden of the fairness of the game',[86] if his own task is limited to that of *understanding* history, without taking part in it, withdrawing into an ivory tower and avoiding any commitment to a political life or to a party.

Obviously, an intellectual is also a man of his own times, somehow influenced by the ideas and values then current. If as a citizen he can serve a political party, as an historian he has no other task, or duty, than that of providing for the advancement of science, of *knowledge*, of democracy, of liberty, of a superior form of civilization.[87] Descending into the arena of a political party, means serving the interests of one's own party, betraying in this way the true nature of science, that is knowledge.

Conclusion of this section

The conclusion is that Mommsen, from a general, philosophical point of view, was far from being a *liberal*, in the sense of *intelligĕre*, in the sense of *understanding* what society is. He took part in political life of his time as a citizen, but lacked any superior consciousness of the concept of what society, as a *whole*, is. He was unaware of what the philosophy of liberalism was, from Locke to Tocqueville. Hence his praise of the tyrants, Caesar and Napoleon III, even while, in Germany, he opposed Bismarck and the *ostelbischen Junker*. In the political struggles of his own country he saw in

---

(in the worst sense, of blind ideological fanaticism), i.e. to put oneself from a point of view <critical>, the only one fecund in scientific research", A. Gramsci, *Il Materialismo Storico e la Filosofia di Benedetto Croce*, Einaudi (Torino, 1966), p. 21. Translation by author.

[86] I am echoing again the vocabulary of Antonio Gramsci, one of my favourite authors.

[87] I am here echoing the vocabulary of Benedetto Croce, in his *Manifesto degli Intellettuali Antifascisti*.

opposing parties *Feinde*, enemies, not people who, with their contrasting views, contributed to the whole. He was an actor on the stage of history, not a superior mind able to conceive the whole, or *Gesamtheit*. That, by contrast, is exactly what Cicero was, with his concept of the mixed constitution.

Certainly, Mommsen was a great philologist and historian, and his prose captivates the reader with its vigorous style but, as for general, philosophical ideas, he is completely unaware of what philosophy of liberalism is, and substantially deaf to what the world of ideas is.

The nineteenth century produced in France Constant and Tocqueville, but the eighteenth century had already produced Montesquieu and, in Scotland, Adam Smith and Adam Ferguson, the last having been the master of Constant in Edinburgh. In relation to political ideas the most relevant issue in eighteenth century had been the controversy about *legislators*, or *lawgivers*, essentially as an answer to Hobbes and Rousseau.

To understand liberalism, and eighteenth-century liberalism in particular, the controversy can be briefly summarized with the words of Adam Smith and these of Adam Ferguson, *der alte Adam, the master of Adam Smith*, as Karl Marx calls him, erroneously considering him as a precursor of philosophical materialism.

Smith wrote, in his *Theory of Moral Sentiments*:

> "The man of system ... is often so enamored with the supposed beauty of his own ideal plan of government, that he cannot suffer the smallest deviation from any part of it ... to insist upon establishing, and upon establishing all at once, in spite of all opposition, everything which that idea may seem to require, must often be the highest degree of arrogance. It is to erect his own judgment into the supreme standard of right and wrong ... It is upon this account, that of all political speculators, sovereign princes are by far the most dangerous. This arrogance is perfectly familiar to them. They entertain no doubt of the immense superiority of their own judgment".[88]

---

[88]Smith, *Theory of Moral Sentiments*, op. cit., VI, ii, 2., & 17-18, pp. 233-4.

As for Adam Ferguson, one cannot understand the true nature of his work, of his *Essay on the History of Civil Society*, without emphasizing his controversy against the *projectors, designing men, men of speculation, men of system, visionary plans, schemes of perfection, general reasoners*, etc.

He writes as follows:

> "Men of speculation have in vain endeavoured to fix a model of government adapted to all mankind";[89] "speculative men have retarded the progress of Science";[90] "free institutions never take rise from the scheme of any single projector";[91] "the best projectors can do is to be cautious of hurting an interest they cannot greatly promote ... nature has not entrusted the preservation of her works to their visionary plans";[92] "the forms of society are derived from an obscure and distant origin; they arise, long before the date of philosophy, from the instincts, not from the speculations, of men ... Every step and every movement of the multitude, even in what are termed enlightened ages, are made with equal blindness to the future; and nations stumble upon establishments, which are indeed the result of human action, but not the execution of any human design ... (societies) admit of greatest revolutions when no change is intended, and ... the most refined politicians do not always know whither they are leading the state by their projects".[93]

These are, briefly sketched, the fundamental concepts of the philosophy of liberalism, which so clearly contrast not only with despotism, but also with *enlightened despotism* and authoritarian democracy. And we do not need to repeat that they impressively coincide with Cicero's statements.

---

[89] Ferguson, *Institutes of Moral Philosophy* (Edinburgh, 1785), p. 288.

[90] Ibid, p. 10.

[91] Ferguson, *An Essay on the History of Civil Society*, ed. D. Forbes (Edinburgh, 1966), p. 134.

[92] *An Essay*, pp. 140-1.

[93] *An Essay*, Part III, sect. ii, p. 122. Friedrich von Hayek titles after Ferguson 'The Result of Human Action but not of Human Design' chap. 6 of his *Studies in Philosophy, Politics and Economics*, Routledge and Kegan Paul (London, 1967).

Mommsen, nevertheless, believed in the *legislator*, in Caesar and in Napoleon III, not in the natural movements of society, not in historical change as a philosophical category, and in the concept of society itself. From his observatory, which was Germany of 1848, he could scarcely understand the true nature of modern liberal societies and the strife of democracy. To understand these, one must think of Smith's *merchant and manufacturer* who changed the face of European society in eighteenth century, and of the American Revolution, which established the principle of democratic liberties, of a self governing people, that had left behind in Europe the traditions of monarchy and the principle of authority. And Germany was at the centre of that old part of the European world that only in the middle of the nineteenth century was able to fight for the principle of constitutional monarchies and for the independence of nations.

See, for example, the *Grundrechte des deutschen Volkes*, for which Mommsen wrote a commentary for the publishers Wigand in 1849. The author's commentary is obviously liberal in a political sense, but the problems he deals with give, especially in the following section, an exact picture of what Germany was at the time:

> "Beware of the false prophets, who speak of absolute equality, equality of the wise and of the silly, of the poor and of the rich, of the rogues and of the honest citizen: it isn't this equality, which throws through the window all the differences natural and necessary, and that is in reality the greatest inequality, because God did not make men all equal, on the contrary He made them very different from each other, and everybody will be judged according to his own merits, it isn't this the equality that you can claim, rather it is equality before the law".[94]

These concepts can well be inscribed in the history of European liberalism, but only marginally. We can well say, with Montesquieu, that "la liberté est le droit de faire tout ce que les lois permettent."[95]

---

[94] See Italian edition, Mommsen, *I diritti fondamentali del popolo tedesco. Commento alla Costituzione del 1848*, a cura di G. Valera, Il Mulino (Bologna, 1994). Translation by author.

[95] Montesquieu, *Esprit des Lois*, book XI, chap. iii.

Nevertheless Montesquieu was writing a century before, under the despotism of French monarchy. The aim of Mommsen seems to be here rather to restrain the movement of liberation, especially of social liberation. He wrote, no doubt, from the point of view of the middle classes of early nineteenth-century Germany.

Therefore, it is no surprise that for him Caesar is *das schöpferische Genie*, "the creator genius",[96] *der Realist und Verstandesmensch*, "the realist and rational person",[97] *der Staatsmann im tiefsten Sinne des Wortes*, "the statesman in the most profound meaning of the word",[98] substantially the highest point of Roman History. Mommsen portrayed his 'democratic monarchy' as the *Vertretung der Nation durch ihren höchsten und unumschränkten Vertrauensmann*, "representation of the nation by the man who, in the highest and most unlimited sense, enjoys its confidence".[99]

As for Napoleon III, he was fascinated by his ideology, and saw him *als Vorbild eines rechten Staatsmanns*, "as the model of a real statesman". In May 1863 he wrote from Paris to his wife the following words: "He has definitely given me the impression of being an eminent man, such that we should have him in our nation".[100]

Nevertheless his opinions and argument about political problems were often contradictory, and no wonder that "a recurring topos of the argument towards Mommsen was that, each hour that he spent with politics instead of science, was an irreplaceable loss for science of antiquity".[101] Politics was not so important.

---

[96] Mommsen, *Römische Geschichte*, III, 461.

[97] Ibid., III, 463.

[98] Ibid.

[99] Ibid, III, 476. He adds, in *Römisches Staats-Recht*, vol. II, p. 755: "Since the principle of the monarchy leads by logical sequence either from its religious side up to the king-god, or from its legal side to the king-master, we must recognise in this procedure that absolute and unshrinking thoroughness of thought and action, which, here as elsewhere, vindicates for Caesar a unique station in history". This citation is from Strachan-Davidson, *Cicero and the Fall of the Roman Republic* (New York-London, 1902), p. 377 n.

[100] "er mir durchaus den Eindruck eines bedeutenden Mannes gemacht hat, wie man ihn unserer Nation wohl wünschen möchte", Rebenich, op. cit., p. 95.

[101] "Ein immer wiederkehrender Topos der gegen Mommsen gerichteten Polemik ... Jede Stunde, die M. statt mit der Wissenschaft mit der *Politisieren* zubrachte, wurde zu einem unersetzlicher Verlust für die Altertumsforschung erklärt...", Rebenich, op. cit., p. 191.

"Für die liberale Reforme der Reichsverfassung agitierte er selten", "he hardly ever was agitated for the liberal reformation of the Constitution", admits the most recent and most informed of his biographers.[102] When visiting Sicily "the social critic of J. G. Seume ... was alien to him. What enraged him was not the exploitation of the country by the clergy and the nobility, but the bad state of preservation of some inscriptions and the stench of the pits of disulphide".[103]

And, last, in his *Agli Italiani*, in 1870,[104] he proved to be scarcely credible in his reasoning in favour of the friendship between Prussia and Italy against France.

In fact Mommsen insisted that, in 1866, after the victory at the battle of Sadowa, Prussia was loyal to the Italians, who had been defeated by the Austrians at Custoza. He admitted that, in 1859, while the Italians were fighting for their own independence with French help, Prussia was ready to support Austria to re-conquer Milan and Northern Italy. The liberals of the North were nevertheless favourable to Italian independence, and he complained that

> "purtroppo è vero, che per lungo tempo, la Prussia ha servito da manutengolo alla politica di Schoenbrunn",
> "it is unfortunately true that for a long time Prussia served as an accomplice to the politics of the Austrian Court".[105]

The point is that the Italians were suspicious of German unification, much more than they were of the policy of Napoleon III in favour of the Catholic Church. And Terenzio Mamiani, the influential philosopher and politician, as Mommsen himself admitted,

> "trova nella unificazione della grande nazione germanica una minaccia continua diretta contro tutti gli altri

---

[102] Rebenich, op. cit., p. 169.

[103] "Die Sozialkritik eines Johann Gottfried Seume (s. sein *Spaziergang nach Syrakus*, 1803) war ihm fremd. Auf Sizilien ärgerte er sich nicht über die Ausbeutung des Landes durch Klerus und Adel, sondern über den schlechten Erhaltungszustand einiger Inschriften und den Gestank von Schwefelgruben", Rebenich, op. cit., pp. 51-2.

[104] See Mommsen, *Lettera agli Italiani* (Firenze, 1870).

[105] Ibid. p. 7.

stati", "finds in the unification of the great German nation a continuous threat levelled against all the other
states".[106]

In 1841 the construction of the 46-meter-high statue of Arminius
had also started, the German national hero, who ambushed and
slaughtered the Roman legions at Kalkriese, in Teutoburger Wald,
Lower Saxony. The construction of the statue was completed in
1875, with the financial help of the state of Prussia. The ambush
of Kalkriese prevented the Roman conquest of Germany, and set
the conditions for the division of Western Europe into two cultural
worlds and into two traditions, mostly hostile to each other in the
centuries to come.

Therefore Mommsen, by nature and constitution unable to understand the profound meaning of Cicero's words, should be now
buried with all honours. The towering figure of Cicero is in fact too
substantial to be liable to any criticism on the part of scholars who
are incapable of a proper moral understanding, their own judgment
being clouded, as it is, by political passions.

But it is now time to leave Mommsen and to return to Cicero.

Section II: Ciceronianism, Newtonianism and Eighteenth-Century Cosmology

What is alive and what is dead in the philosophy of Cicero[107]

For the followers of Mommsen -there are, or there were, still a
few[108]- not much is alive of Cicero's philosophy, while, by contrast,
much is still alive for the wider world of learning, as we shall demonstrate.

The point is whether philosophical research can today boast of
having realized a *progress* in appreciating Ciceronianism. It has,

---

[106]Ibid., p. 19.

[107]I am echoing the title of Benedetto Croce's essay *Ció che è vivo e ció che è morto della filosofia di Hegel.*

[108] For example Andreas Alföldi (Budapest 1895-Princeton 1981), who was fascinated by the personality of Caesar and disappointed by what he considered
the 'opportunism' of Cicero, whom he accuses of having been the ideological
instigator of Caesar's assassination.

obviously. Nevertheless, new philosophical systems, or new philoso-phies, need time to be elaborated, because they represent reflection upon centuries of history.

Plato and Aristotle expounded all the ideas that the science of antiquity was able to theorize, and Cicero summarized, and largely re-elaborated these ideas with a Roman flavour of his own, in what we can call his *philosophical encyclopaedia*, bringing them to the threshold of Christianity. In this sense he made a substantial ad-vance on them. The next step was represented by St. Augustine's mysticism, while St. Thomas is the philosopher of the Middle Ages, and his system, which is a reworking of Aristotelianism, was super-seded only by Galileo and Newton. Then came the philosophy of Enlightenment, with its particular features which, in the principal European nations, were obviously the exact mirror of the progress of these societies. With the age of Enlightenment, especially with the Scottish one, Ciceronianism still had something to say. Only with Kant and Hegel was it substantially superseded, supposing that the 'eternal problems of philosophy' are ever superseded.

We don't need here to write a history of Ciceronianism, because this task has been remarkably well accomplished by Thäddäeus Zielinski in particular, and by more recent, talented historians.[109] Nevertheless we believe that something needs to be added, both on Newtonianism and on eighteenth-century cosmology, with its opti-mism and the concept of the best of all possible worlds.

The 'eternal problem of philosophy', the existence of God the Creator, as in Hegel's *Phenomenology* or, alternatively, the exist-ence of the original Chaos, as in materialistic philosophy, can be here summarized in the words of Friedrich Engels:

> "The great basic question of all, and especially of re-
> cent philosophy, is that of the relation of the Thought
> to Being ... of the Spirit to Nature, the highest question
> of all philosophy ... that, after all, played a great role

---

[109]See his *Cicero im Wandel der Jahrhunderte*, Sechste Auflage (Stuttgart, 1973), Nachdruck der 3. Auflage (Leipzig, 1912). See furthermore Matthew Fox, *Cicero's Philosophy of History*, Clarendon Press (Oxford, 2007), and C. Steel ed., *The Cambridge Companion to Cicero* (Cambridge UP, 2013), particularly part III, 'Receptions of Cicero'. The chapter on the eighteenth century, ibid., pp. 318-336, is by Matthew Fox.

also in the Scholastics of the Middle Ages, the question: what was there in the beginning, Spirit or Nature? The question, facing the Church, amounted to this: did God create the World, or did the World exist from Eternity? According to how this question was answered, philosophers divided themselves in two great camps. Those who believed that originally there was Spirit rather than Nature, and also, in the last resort, somehow believed in the Creation of the World ... formed the camp of Idealism. The others, who at the origin saw Nature, belong to the different Schools of Materialism."[110]

This problem, obviously, is commonly debated in philosophical speculation, and being unique it will be debated for centuries, or millennia to come.

Nevertheless it represents no more than the problem of materialism of antiquity, i.e. of Epicureanism, and at the same time of its opposing school, Stoicism, and this is, obviously, the fundamental question debated in Cicero's philosophy.

Newton and Ciceronianism

Ciceronianism, no doubt, is the fundamental chapter in European cultural heritage. In the words of Walter Rüegg, commenting on Dilthey, Cicero's writings "appear as having exerted the strongest

---

[110]Translation by author. The German text reads as follows: "Die große Grundfrage aller, speziell neueren Philosophie, ist die nach dem Verhältnis des Denkens zum Sein ... des Geistes zur Natur, die höchste Frage der gesamten Philosophie ... die übrigens auch in der Scholastik des Mittelalters ihre große Rolle gespielte, die Frage: was ist das Ursprüngliche, der Geist oder die Natur? ... Diese Frage spitzte sich, der Kirche gegenüber, dahin zu: hat Gott die Welt erschaffen, oder ist die Welt von Ewigkeit da? Je nachdem diese Frage so oder so beantwortet wurde, spalteten sich die Philosophen in zwei große Lager. Diejenigen, die die Ursprünglichkeit des Geistes gegenüber der Natur behaupteten, also in letzter Instanz einer Weltschöpfung irgendeiner Art annahmen ... bildeten das Lager des Idealismus. Die andern, die die Natur als das Ursprüngliche ansahen, gehören zu den verschiedenen Schulen des Materialismus ...", Friedrich Engels, *Ludwig Feuerbach und der Ausgang der klassischen deutschen Philosophie*, in Marx-Engels *Werke*, Dietz Verlag (Berlin, 1962), Band 21, pp. 274-5.

influence in the shaping of the European spiritual world, along with the Bible".[111]

And the philosophy of the Stoics, which is substantially Cicero's philosophy, can be summarized in his own words.

As for pronoia, or providence, it is the fundamental concept in the *De Natura Deorum*.[112] Hence the concepts that are inferred from it.

As for finalism or teleology in nature:

> "... it is undeniable that every organic whole must have an ultimate ideal of perfection. As in vines or in cattle we see that, unless obstructed by some force, nature progresses on a certain path of her own to her goal of full development ... in the world of nature as a whole there must be a process towards completeness and perfection ... it follows of necessity that the world is an intelligent being, and indeed also a wise being.

> Again, what can be more illogical than to deny that the being which embraces all things must be the best of all things or, admitting this, to deny that it must be, first, possessed of life, secondly, rational and intelligent, and lastly endowed with wisdom? ... there is nothing else beside the world that has nothing wanting, but is fully equipped and complete and perfect in all his details and parts ... man himself however came into existence for the purpose of contemplating and imitating the world; he is by no means perfect, but he is 'a small fragment of that which is perfect'. The world, on the contrary, since it embraces all things and since nothing exists which is

---

[111] See W. Dilthey, *Gesammelte Schriften*, Dritte Unveränderte Auflage (Leipzig und Berlin 1929), 1914, II, p. 499: "Diese Überzeugungen (concerning metaphysics) sind auch nicht vom Christentum hervorgebracht oder durch irgendeine Religion hergestellt. Wir finden sie in den Werken des Cicero, die den gebildeten Bewohnern der römischen Provinzen das Ergebnis des Altertums vermittelten, ebenso ausgedrückt als später in den Schriften der Kirchenväter". And Walter Rüegg, commenting on the words of Dilthey, adds that Cicero was "eine Persönlichkeit, deren Werk neben der Bibel den größten Anteil an der Gestaltung der europäischen Geisteswelt zu besitzen scheint", W. Rüegg, *Cicero und der Humanismus*, op. cit., Vorwort, S. VII.

[112] It is demonstrated in II, 73-153.

not within it, is entirely perfect; how then can it fail to possess that which is the best? But there is nothing better than intelligence and reason; the world therefore cannot fail to possess them";[113]

"... it would have been the proper course for the philosophers, if it so happened that the first sight of the world perplexed them, afterwards when they had seen its definite and regular motions and all its phenomena controlled by fixed system and unchanging uniformity, to infer the presence not merely of an inhabitant of this celestial and divine abode, but also of a ruler and governor, the architect as it were of this mighty and monumental structure".[114]

As for cosmology:

"Who ... on seeing the regular motions of the heaven and the fixed order of the stars and the accurate interconnection and interrelation of all things, can deny that these things possess any rational design ...? When we see something moved by *machinery*, like an *orrery* or *clock* or many other such things, we do not doubt that these contrivances are the works of reason; when therefore we behold *the whole compass of the heaven moving with revolutions of marvellous velocity and executing with perfect regularity the annual changes of the*

---

[113] "Neque enim dici potest in ulla rerum institutione non esse aliquid extremum atque perfectum ... sic in omni natura ac multo etiam magis necesse est absolvi aliquid ac perfici ... necesse est intelligentem esse mundum et quidem etiam sapientem ... ipse autem homo ortus est ad mundum contemplandum et imitandum, nullo modo perfectus, sed est quaedam particula perfecti. Sed mundus quoniam omnia conplexus est neque est quicquam quod non insit in eo, perfectus undique est; quid igitur potest ei deesse id quod est optimum? Nihil autem est mente et ratione melius; ergo haec mundo deesse non possunt." Cicero, *De Natura Deorum*, II, xii-xiv.35-38.

[114] "...sic philosophi debuerunt, si forte eos primus aspectus mundi conturbaverat, postea, cum vidisset motus eius finitos et aequabiles omniaque ratis ordinibus moderata immutabilique constantia, intellegere inesse aliquem non solum habitatorem in hac caelesti ac divina domo sed etiam rectorem et moderatorem et tamquam architectum tanti operis tantique muneris", *De Natura Deorum*, II, xxxv, 90.

> *seasons with absolute safety and security* for all things,
> how can we doubt that all this is effected not merely
> by reason, but by a reason that is transcendent and di-
> vine?".[115]

As for the existence of a transcendent God:

> "For when we gaze upward to the sky and contemplate
> the heavenly bodies, what can be so obvious and so
> manifest as that there must exist some power possess-
> ing transcendent intelligence by whom these things are
> ruled?".[116]

As for optimism, or the concept that this is the best of all possible worlds, which gives the clue to understand the conception of
harmonies in eighteenth-century philosophy, it is clearly stated in
Cicero's sentences:

> "Now the government of the world contains nothing that
> could possibly be censured; given the existing elements,
> the best that could be produced from them has been
> produced. Let someone prove that it could have been
> better. But no one will ever prove this".[117]

As for the force of gravity:

---

[115] "Quis enim hunc hominem dixerit qui, cum tam certos caeli motus tam ratos
astrorum ordines tamque inter se omnia conexa et apta viderit, neget in his
ullam inesse rationem, eaque casu fieri dicat quae quanto consilio gerantur
nullo consilio adsequi possumus? An, cum machinatione quadam moveri
aliquid videmus, ut sphaeram ut horas ut alia permulta, non dubitamus
quin illa opera sint rationis, cum autem impetum caeli cum admirabili celeri-
tate moveri vertique videamus constantissime conficientem vicissitudines an-
niversarias cum summa salute et conservatione rerum omnium, dubitamus
quin ea non solum ratione fiant sed etiam excellenti divinaque ratione?", *De
Natura Deorum*, II.xxxviii.97.

[116] "Quid enim potest esse tam apertum tamque perspicuum, cum caelum sus-
peximus caelestiaque contemplati sumus, quam esse aliquod numen praes-
tantissimae mentis quo haec regantur?", *De Natura Deorum*, II, ii, 4.

[117] "Cuius quidem administratio nihil habet in se quod reprehendi possit; ex
iis enim naturis quae erant quod effici optimum potuit effectum est. Do-
ceat ergo aliquis potuisse melius; sed nemo umquam docebit", *De Natura
Deorum*, II.xxxiv.86-7.

> "That rational and intelligent substance ... which draws and collects the outermost particles towards the centre ... all its parts must converge towards the centre ... its vast complex of gravitational forces ... on the same principle the sea ... seeks the earth's centre and so is massed into a sphere uniform on all sides ... the stars ... maintain their spherical form by their own internal gravitation ...".[118]

Cleanthes as speaking in Cicero:

> "Cleanthes said ... the uniform motion and revolution of the heavens, and the varied groupings and ordered beauty of the sun, moon and stars, the very sight of which was in itself enough to prove that these things are not the mere effect of chance ... he realizes that there is someone who presides and controls ... these mighty world-motions are regulated by some Mind ... therefore God does exist ... nothing exists among all things that is superior to the world, nothing that is more excellent or more beautiful; and not merely does nothing superior to it exist, but nothing superior can even be conceived ... These processes and this musical harmony of all the parts of the world assuredly could not go on

---

[118] "Maxime autem corpora inter se iuncta permanent cum quasi quodam vinculo circumdato colligantur; quod facit ea natura quae per omnem mundum omnia mente et ratione conficiens funditur et ad mediam rapit et convertit extrema. Quocirca si mundus globosus est ob eamque causam omnes eius partes undique aequabiles ipsae per se atque inter se continentur, contingere idem terrae necesse est, ut omnibus eius partibus in medium vergentibus (id autem medium infimum in sphaera est) nihil interrumpat quo labefactari possit tanta contentio gravitatis et ponderum", *De Natura Deorum*, II, xlv.115.116.

On Newton and the force of gravity, Pemberton wrote: "In 1665 he fell into a speculation on the power of gravity ... that this power must extend much further than is usually thought. Why not as high as the moon? said he to himself, and, if so, her motion must be influenced by it: perhaps she is retained in her orbit thereby", H. Pemberton, *A View of Sir Isaac Newton's Philosophy* (1728), Preface.

were they not maintained in unison by a single divine and all-pervading spirit".[119]

Newtonianism

Newton's physics, as expounded in the *Principia* of 1687, was presented in a popular form by Richard Bentley (1662-1742) in his Boyle lectures. We need to transcribe here some sections from them:

> "... the Divine inspection into the affairs of the World doth necessarily follow from the Nature and Being of God. And he that denies this, doth implicitly deny his Existence ... in his heart he hath said, There is no God. A God, therefore a Providence; was a general argument of vertuous men, and not peculiar to the Stoicks alone. And again, No Providence, therefore no God, was the most plausible reason, and the most frequent in the mouths of Atheistical men ... There are some Infidels among us, that not only disbelieve the Christian Religion; but impugn the assertion of a Providence, of the Immortality of the Soul, of an Universal Judgment to come, and of any Incorporeal Essence; and yet to avoid the odious name of Atheists, would shelter and skreen themselves under a new one of Deists, which is not quite so obnoxious ... The Fool, that doth exempt the affairs of the world from the ordination and disposal of God, hath said in his heart, There is no God at all";[120]

---

[119] "Quartam causam esse eamque vel maximam aequabilitatem motus conversionumque caeli, solis lunae siderumque omnium distinctionem varietatem pulchritudinem ordinem, quarum rerum aspectus ipse satis indicaret non esse ea fortuita.....cum videat omnium rerum rationem modum disciplinam non possit ea sine causa fieri iudicare sed esse aliquem intellegat qui praesit et cui pareatur ... statuat necesse est ab aliqua mente tantos naturae motus gubernari ... Etenim si di non sunt, quid esse potest in rerum natura homine melius? In eo enim solo est ratio, qua nihil potest esse praestantius; esse autem hominem qui nihil in omni mundo melius esse quam se putet desipientis adrogantiae est; ergo est aliquid melius; est igitur profecto deus ... Haec ita fieri omnibus inter se concinentibus mundi partibus profecto non possent nisi ea uno divino et continuato spiritu continerentur", *De Natura Deorum*, II, vi.15-19.

[120] Richard Bentley, *The Folly of Atheism, and (what is now called) Deism. A Sermon Preached at Saint Martin's in the Fields, March 7th 1692*. Being

"a noble saying of the Emperor Marcus, That he would not endure to live one day in the world, if he did not believe it to be under the Government of Providence";[121]

"... the Epicureans, who maintained that the world was owing to the fortuitous concourse of Atoms... the Peripateticks, that supposed all things to have been eternally, as they now are, and never to have been made at all, either by the Deity or without him";[122]

"There is ... an Immaterial and Intelligent Being that created our Souls: Which Being was either Eternal of it self, or Created Immediately or ultimately by some other Eternal Being, that hath all those Perfections. There is therefore originally an Eternal, Immaterial, Intelligent Creator, all of which together are the Attributes of God alone".[123]

Then Bentley proceeds to 'A Confutation of Atheism from the Structure and Origin of Humane Bodies', closely following Ciceronian arguments:

"... the visible marks of Divine Wisdom ... the Organical Structure of Human Bodies, whereby they are fitted to live, and move, and be vitally informed by the Soul, is unquestionably the workmanship of a most wise, and powerfull, and beneficent Maker";[124]

"Bodies of Men ... are excellently well fit for Life, and Motion, and Sensation ... The Eye is very proper and meet for seeing, the Tongue for tasting and speaking, the Hand for holding and lifting ... as the effect of Contrivance and Skill, and consequently the workmanship of a most Intelligent and Beneficent Being";[125]

the First of the Lecture Founded by the Honourable Robert Boyle, Esquire, by Richard Bentley, M.A. The Third Edition (London, 1692), ibid., p. 5.

[121] Bentley, ibid., part 1, p. 12.

[122] Bentley, ibid., part 2, 'Matter and Motion', pp. 6-7.

[123] Bentley, ibid., part 2, p. 31.

[124] Bentley, 'A Confutation', 3, part 1, pp. 6-8.

[125] Bentley, ibid., pp. 8-9.

"Something, we are sure, must have Existed from all Eternity, because all things could not emerge and start out of Nothing ... Some Being, though infinitely above our finite comprehension, must have had an identical, invariable Continuance from all Eternity; which Being is no other than God. For all his Nature is perfect and immutable without the least shadow of change; so his Eternal Duration is permanent and indivisible, not measurable by Time and Motion, nor to be computed by number of successive moments".[126]

In part I Bentley confuted Aristotelian atheists who, "to avoid the difficulties of the first production of Mankind, without the intervention of Almighty Wisdom and Power, will have the race to have thus continued without beginning by an eternal succession of infinite past Generations, and the notion of the Astrological undertakers, that would raise Men like Vegetables out of some fat and slimy soil well digested by the kindly heat of the Sun, and impregnated with the influence of the Stars upon some remarkable and periodical conjunctions: Which opinion had been vamp'd up of late by Cardan and Cesalpinus ... Other Atheists resolve the whole business into the unaccountable shuffles and tumults of Matter, which they call Chance and Accident. The production of Humane Bodies by Mechanism and Necessity ... the mechanical or corpuscolar philosophy ... If we consider the Phenomena of that Material World with a due and serious attention, we shall plainly perceive, that its present frame and constitution and the established Laws of Nature are constituted and preserved by Gravitation alone. That is the powerfull cement, which holds together this magnificent structure of the world; which stretched the North over the empty space, and hangeth the Earth upon Nothing ... Without that the whole Universe, if we suppose an undermin'd power of Motion infused into Matter, would have been a confused Chaos, without beauty or order ... this Gravity, the great Basis of all Mechanism, is not it self Mechanical; but the immediate Fiat and Finger of God, and the Execution of the Divine Law, and that the Bodies have not the power of tending towards a Centre, either from other Bodies or from themselves ... No Com-

---

[126]Bentley, ibid., p. 20.

pound Body in the visible world can subsist and continue without Gravity, and Gravity do immediately flow from a Divine Power and Energy";[127]

> "the Order and Beauty of the Inanimate Parts of the World, the discernible Ends and Final Causes of them .... it is the Product and Workmanship, not of blind Mechanism or blinder Chance; but of an Intelligent and Benign Agent, who by his excellent Wisdom made the Heavens and Earth: and gives Rains and fruitful Seasons for the service of Man".[128]

Isaac Newton, in his turn, answered Bentley in his *Letters* to Richard Bentley:

> "How the Matter should divide itself into two sorts ... I do not think explicable by mere natural causes, but am forced to Ascribe it to the counsel and contrivance of a voluntary Agent ... the motions, which the planets now have, could not spring from any natural cause alone, but were impressed by an intelligent Agent ... there is no natural cause which could determine all the planets ... to move the same way and in the same plane, without any considerable variation: this must have been the effect of counsel ... In the inclination of the Axis, which causes Winter and Summer, and makes the Earth habitable; the diuturnal rotations of sun and planets, as they could hardly arise from any cause purely mechanic-al ... they seem to make up that harmony in the system, which ... was the effect of choice rather than chance";[129]

> "The force of gravity may put the planets into motion, but without the Divine Power it could never put them

---

[127] Bentley, ibid., 'A Confutation of Atheism from the Structure and Origin of Humane Bodies', Part II, pp. 3-6.

[128] Bentley, ibid., 'A Confutation of Atheism from the Origin and Frame of the World', Part I, p. 20.

[129] I. Newton, 'Four Letters to Doctor Bentley; containing some Arguments in Proof of a Deity', in *Opera quae extant Omnia*, (Londinii 1782), Tom. IV, pp. 427-442, ibid., Letter First, pp. 430-2.

into such a circulating motion ... and therefore ... I am compelled to ascribe the frame of this system to an intelligent Agent";[130]

"... the hypothesis of matter's being at first evenly spread through the heavens, is, in my opinion, inconsistent with the hypothesis of innate gravity, without a supernatural power to reconcile them; and therefore it infers a Deity".[131]

Cicero had already summarized these problems with the following words in the *De Republica*:

"For that which is always in motion is eternal, but that which communicates motion to something else, but is itself moved by another force, necessarily ceases to live when this motion ends. Therefore only that which moves itself never ceases its motion ... nay, it is the source and first cause of motion in all other things that are moved. But this first cause has itself no beginning, for everything originates from the first cause, while it can never originate from anything else. And since it never had a beginning, it will never have an end. For if a first cause were destroyed, it could never be reborn from anything else, nor could it bring anything else into being; since everything must originate from a first cause. Thus it follows that motion begins with that which is moved of itself; but this can neither be born nor die".[132]

---

[130] Newton, ibid., Letter II, pp. 436-7.

[131] Newton, ibid., Letter IV, p. 440.

[132] "Nam quod semper movetur, aeternum est; quod autem motum adfert alicui, quodque ipsum agitatur aliunde, quando finem habet motus, vivendi finem habeat necesse est. Solum igitur, quod sese movet, quia numquam deseritur a se, numquam ne moveri quidem desinit; quin etiam ceteris, quae moventur, hic fons, hoc principium est movendi. Principii autem nulla est origo; nam ex principio oriuntur omnia, ipsun autem nulla ex re alia nasci potest; nec enim esset id principium, quod gigneretur aliunde; quodsi numquam oritur, ne occidit quidem umquam. Nam principium exstinctum nec ipsum ab alio renascetur nec ex se aliud creabit, siquidem necesse est a principio oriri omnia. Ita fit, ut motus principium ex eo sit, quod ipsum a se movetur; id autem nec nasci potest nec mori; vel concidat omne caelum omnisque

And in the *Academica*:

"... this world is wise and possessed of an intelligence that constructed both itself and the world, and that controls, moves and rules the universe".[133]

Eighteenth-century cosmology

Just by transcribing texts, which need no comment on our part, we have seen how Newton's arguments are substantially Cicero's ones.[134] And not only in the seventeenth, but even in the eighteenth century Stoic cosmology had not been superseded by any

---

natura et consistat necesse est nec vim ullam nanciscatur, qua a primo impulsa moveatur", Cicero, *De Rep.*, VI, xxv, 27.

Cicero takes this section from *Phaedrus* 245- C-E. B.J.T. Dobbs, 'Newton and Stoicism', *The Southern Journal of Philosophy*, 1985, vol. XXIII, Supplement, pp. 109-123, writes that Newton owned Cicero's *Opera Omnia*, but apparently not Plato's *Phaedrus*. In any case he certainly read Latin well, not so well Greek, if any.

[133] "... mundum esse sapientem, habere mentem quae et se et ipsum fabricata sit et omnia moderetur moveat regat", *Academica*, II, xxvii-xxviii, 118-119.

[134] One could observe that in Newton's system of ideas Ciceronianism did not have the relevance that it had in other authors. In Newton's works reflection about philosophical problems is in fact limited. He "was not a philosopher ... his contributions to knowledge ... were to science, not to philosophy. By contrast, Galileo ... turned the issue of the epistemic authority of theology versus the epistemic authority of empirical science into a hallmark of modern times ... Newton wrote virtually nothing critical of the Aristotelian tradition ... from his extant writings alone, Newton has more claim to being a major theologian than a major philosopher", I.B. Cohen-G.E Smith, *The Cambridge Companion to Newton* (Cambridge, 2002), Introd., p. 1.

Newton spent much of his time in the study of Biblical Chronology, and looked for the philosophical stone. This is revealing in itself, and proves a singular backwardness in philosophical concepts. Behind Galileo's (1564-1642) system there was the Italian Renaissance, Newton's (1642-1727) England was that of the civil wars, of the Restauration, and, last in the century, of the Glorious Revolution. It was only the next century that it knew the intellectual progress of Enlightenment. Ciceronianism, that had not disappeared in the seventeenth century, as the words of Richard Bentley prove, reappeared then in all its force.

Secondary literature repeats the story of Newton's having had the idea of gravitation when seeing an apple falling in a garden etc. etc. Richard S. Westfall writes that "the story does not survive comparison with the record of his early work in mechanics. The story vulgarizes universal gravitation by treating it as a bright idea .... What after all was in the paper that revealed the inverse-square relation? Certainly not the idea of universal gravitation

new conception of the universe. As for the philosophical systems then current, Hume observed in particular that

> "the most durable, as well as justest fame, has been acquired by the easy philosophy, and ... abstract reasoners seem hitherto to have enjoyed only a momentary reputation, from the caprice or ignorance of their own age ... The fame of Cicero flourishes at present, but that of Aristotle is utterly decayed".[135]

The point is that Cicero had still much to say to the spiritual and scientific life of that century. Not so Aristotle, whose system had been superseded by Galileo and Newton, and scarcely entered into the philosophical debate. Particularly in Scotland, as William Whiston (1667-1752) tells us in his *Memoirs*, the philosophy of Newton was flourishing, -that is, we maintain, substantially that of Cicero, to which Newton had added scientific reasons,- while "we at Cambridge ... used to study with ignominy the false hypotheses of Cartesian philosophy".[136]

... to Newton the mechanical philosopher an attraction at a distance was inadmissible in any case ... Newton must have had something in mind when he compared the moon's centrifugal force with gravity, and there is every reason to believe that the fall of an apple gave rise to it", Westfall, *Never at Rest: A Biography of Isaac Newton* (Cambridge UP, 1980), p. 155. The idea of gravitation, as we have seen above (see text corresp. with no. 122), had been theorized by Cicero. Did Newton reach it thanks to his "early work in mechanics"? Supposedly, he knew something about Cicero. Nevertheless, in Westfall's biography references to Cicero are of a quite occasional nature, and occur on p. 57, 353, 807, 822. But, can we really suppose that Newton was unaware of Cicero's words about the force of gravity?

[135] Hume, *An Enquiry concerning Human Understanding*, in *Enquiries*, ed. L. A. Selby-Bigge, third edition, Clarendon Press (Oxford, 1975), Section I, p. 7. In a letter to Lord Kames, he drew up an unsympathetic portrait of Cicero, whose first *Philippic* 'is not much admired by the ancients', while in the second 'he gives a full loose to his scurrility', Hume to Lord Kames, 13 June 1742, in *The Letters of David Hume*, I, pp. 41-2.

[136] Cited by Eric G. Forbes, 'Le origini dell'illuminismo scozzese: filosofia, istruzione, scienza', in *Scienza e filosofia scozzese dell'età di Hume*, A. Santucci ed., (Bologna, 1976), ibid., p. 25. In *Hume's Sentiments. Their Ciceronian and French Context*, EUP (Edinburgh, 1982), p. 30, Peter Jones writes that "in Hume's time, orthodox and free-thinker alike felt the challenge of Cicero, and it may be noted that Professor John Pringle devoted his moral philosophy lectures in the University of Edinburgh, in 1740-1, to *De*

Newtonianism was not only anti-Cartesianism. It was so for its philosophical premises, for the method of enquiry, which is anti-rationalistic. In the matter of cosmology, as we have demonstrated above, Newtonianism is no more than Ciceronianism.

In 1728 John Desaguliers (1683-1744), in his *The Newtonian System of the World*, wrote that "The System of the Universe, as taught by Pythagoras, Philolaus, and others of the Ancients, is the same, which was since revived by Copernicus ... and at last demonstrated by Sir Isaac Newton", adding that "all the difference between the modern and ancient system, is only what is added to it since the invention of the telescope".[137]

Soon after Conyers Middleton (1683-1750), in his *Life of Cicero*, added that

> "several of the fundamental principles of the modern philosophy which pass for the original discoveries of these later times, are the revival rather of ancient notions maintained by some of the first philosophers of whom we have any notice in history: as motion of the earth; the antipodes; a vacuum; and a universal gravitation, or attractive quality of matter; which holds the world in its present form and order".[138]

And Colin McLaurin (1698-1746), a pupil and follower of Newton, a mathematical genius of his own age, professor of mathematics at the university of Edinburgh upon Newton's recommendation, in his *Account* described the operations of nature with a vocabulary that closely followed Cicero's terminology:

> "The plain argument for the existence of the Deity ... is from the evident contrivance and fitness of things for one

---

Officiis and the problems of suicide, immortality, the future state, *decorum*, *modestia*, and the being and attributes of God".

[137] On this subject see V. Merolle, 'Introductory Essay' to Ferguson's *The Manuscripts*, particularly pp. xv-xix.
John Desaguliers, in *The Newtonian System of the World* (1728), cited in A.O. Aldridge, 'Shaftesbury and the Deist Manifesto', *Transactions of the American Philosophical Society*, New Series, vol. 41, part 2 (Philadelphia, June 1951), pp. 295-385, ibid., p. 300.

[138] C. Middleton, *Cicero's Life and Letters* (Edinburgh, 1887), p. 305. The book was first published in 1741.

another, which we meet throughout all parts of the universe ... a manifest contrivance immediately suggests a Contriver ... the structure of the eye ... the ear ... The admirable and beautiful structure of things for final causes, exalt our idea of the Contriver: the unity of design shews him to be One. The greatest motions in the system ... suggest his Almighty Power ... The simplicity of the laws that prevail in the world, the excellent disposition of things, in order to obtain the best ends, and the beauty which adorns the works of nature... suggest his consummate Wisdom".[139]

Anthony Ashley Cooper, Earl of Shaftesbury (1671-1713), Francis Hutcheson (1694-1746), Adam Smith (1723-1790) in his *History of Astronomy* and *History of Ancient Physics*, even Hume (1711-1776) in *The Natural History of Religion* and in the *Dialogues Concerning Natural Religion*, repeated not only concepts, but even Cicero's wording, particularly from the *De Natura Deorum*, about the structure of the universe. And one does not need to forget that the *Mens*, or *Universal Mind* of the Stoics, *was no more than the God of incoming Christianity.*
And Cicero, for his part, had declared that

"for any human being in existence to think that there is nothing in the whole world superior to himself would be an insane piece of arrogance; therefore there is something superior to man; therefore God does exist".[140]

On this subject, let us hear from Anthony Ashley Cooper, the <beloved Plato of Europe>, as Herder called him, the luminous

---

[139] Colin McLaurin, *Account of Sir Isaac Newton's Discoveries*, P. Murdoch (London, 1748), p. 381. Along with these authors, William Derham's *Artificial Clockmaker*, 1696), George Cheyne's *Philosophical Principles of Natural Religion* (1705) and, last, William Paley's *Natural Theology* (1809), should be considered. In these works the image of the clockmaker of the universe is recurrent (e.g., G. Cheyne, *Philosophical Principles*, I, IV, p. 5)

[140] "Etenim si di non sunt, quid esse potest in rerum natura homine melius? In eo enim solo est ratio, qua nihil potest esse praestantius; esse autem hominem qui nihil in omni mundo melius esse quam se putet desipientis adrogantiae est; ergo est aliquid melius; est igitur profecto deus", *De Natura Deorum*, II, v, 16-17.

precursor of Enlightenment, where he played a fundamental role in the Deistic movement.

In *The Moralists, a Philosophical Rhapsody*, Shaftesbury praises "the mutual dependence of things! The relation of one to another ... the order, union and coherence of the whole". There is a "just relation of all these parts to one another ... think of the many parts of the vast machine ... having recognised this uniform consistent fabric, and owned the universal system, we must of consequence acknowledge a universal mind ... a God-governed machine".[141]

As for optimism:

> "Every particular nature certainly and constantly pro-
> duces what is good to itself, unless something foreign
> disturbs or hinders it ... every particular nature thrives
> and attains its perfection ... all it produces is to its own
> advantage and good, the good of all in general; and what
> is for the good of all in general is just and good".[142]

Then the stupendous hymns to the "glorious nature", to the "perfection of Being", follow. They are Ciceronian in inspiration, substantially a paraphrasis from Cicero's works, but also let the reader think of Boethius's hymn to the 'Creator of the starry heavens' in book I of his *De Consolatione Philosophiae*.[143]

---

[141] See *The Moralists*,(published 1709), *in Characteristics of Men, Manners, Opinions, Times*, edited by John M. Robertson, The Bobs Merrill Co (Indianapolis, 1964), vol. II, Part II, Sect. V, pp. 65-7.

[142] Ibid., pp. 106-7.

[143] Ibid., Part III, Sect. I, p. 97, p. 110 ff. According to A.O. Aldridge, op. cit., p. 300, not necessarily what we find in the *Characteristics* is Newtonian, because "Pythagorean and Stoic thought were permeated with reasoning from nature to God, and the Shaftesburian cosmology is primarily Stoic ... Shaftesbury probably knew of Newton's theories, yet Cicero or Pythagoras could just as well have inspired the various rhapsodies in the *Moralists* concerning the multitude of fixed stars, visible and invisible..."

See also G. Gawlick, 'Cicero & the Enlightenment', in *Studies on Voltaire*, XXV, 1963, pp. 657-82, that deals with the principal writers of Enlightenment, Montesquieu, S. Clarke, Lord Cherbury, Cudworth, Fontenelle, Toland, Diderot, Molyneaux, Collins, Bentley, Berkeley, Middleton, but the author is interested in deism, not at all in physics or cosmology.

On these problems see R. Wright, *Cosmology in Antiquity*, Routledge (London, 1995).

Let us now hear from Hume.

In the *Natural History of Religion* he speaks of

> "one single being, who bestowed existence and order on this vast machine, and adjusted all its parts, according to one regular plan or connected system ... All things in the universe are evidently of a piece. Every thing is adjusted to every thing. One design prevails throughout the whole. And this uniformity leads the mind to acknowledge one author";[144]

> "a purpose, an intention, a design is evident in every thing ... we must adopt, with the strongest conviction, the idea of some intelligent cause or author... Even the contrarieties of nature, by discovering themselves every where, become proofs of some consistent plan, and establish one single purpose or intention".[145]

In the *Dialogues concerning Natural Religion* he has Cleanthes saying:

> "Look around the world: contemplate the whole and every part of it: You will find it to be but one great machine, subdivided into an infinite number of lesser machines, which again admit of subdivisions, to a degree beyond what human senses and faculties can trace and explain. All these various machines, and even their most minute parts, are adjusted to each other with an accuracy, which ravishes into admiration all men ... The curious adapting of means to ends, throughout all nature, resembles exactly, though it much exceeds, the productions of human contrivance, of human designs, thought, wisdom and intelligence".[146]

---

[144] Hume, *The Natural History of Religion*, Sect. II-Origin of polytheism.

[145] *The Natural History of Religion*, Sect XV, General Corollary.

[146] *Dialogues concerning Natural Religion*, Part II, in *Philosophical Works*, II, pp. 391-2. Concerning Hume, Peter Jones, in his *Hume's Sentiments*, op. cit., complains that "most of Cicero's works are unfamiliar to modern students of Hume", and emphasizes the importance of the letter of 1739 by

For the Scots, therefore, but not less so for the European world
of learning, Ciceronian cosmology was the common, philosophical
heritage. It was theorized with particular clarity by Adam Smith in
his *Essays on Philosophical Subjects*, of which we need to transcribe
here entire sections, to show how their direct source was no more
than in Cicero's works. Nevertheless, while Hume's frame of reasoning is substantially sceptical and anti-religious, far from being
so are the philosophical ideas of Smith, always prudent in matters
of religion, such that he refused to be the literary executor of Hume
for the publication of the *Dialogues concerning Natural Religion*.[147]
Let us now hear from Adam Smith:

"Human society, when we contemplate it in a certain abstract and philosophical light, appears like a great, an

Hume to Hutcheson, in which Hume writes: "You are a great admirer of
Cicero, as well as I am" (HL, I, Letter 13, p. 35).

A main point made by Jones is that "for application of 'experimental' philosophy to moral subjects Hume needed to understand little more of Newton
than the general statements of method in *Principia* and *Opticks*, which were
explained by every commentator ... Hume's lack of interest in science finds
an explanation in his deep commitment to Ciceronian humanism, with its
distinctive attitude to man and to philosophy; such views separated him
from Newton", Jones, p. 42. Furthermore: "Every educated reader could
discern at the time of its posthumous publication, that Hume's *Dialogues
concerning Natural Religion* was modelled on Cicero's *De Natura Deorum*.
Most readers, no doubt, could also discern the Ciceronian influence on the
earlier *Natural History of Religion*, where it was modestly advertised in
footnotes, and in the purely philosophical analysis of theology and religion,
where it was not mentioned at all. The debt to Cicero in *Treatise* Book III, to
which Hume referred in his 1739 letter to Hutcheson ... is not acknowledged
in the text, although references are given in the rewritten version of the
second *Enquiry*", Jones, ibid., p. 29.

The points made out by Jones are the points of the present author.

According to Christopher J. Berry, "there is no systematic account of
Stoicism in the eighteenth century, and one good reason for that is that
it would be almost impossible to write simply because it would have to
incorporate and encompass so much of what was written", Ch. Berry, 'Smith
under Strain', *European Journal of Political Theory*, vol. 3, no. 4, Oct.
2004, pp. 455-64, ibid. p. 457.

[147] See his letter, dated 5 Sept. 1776 to Strahan the publisher: "... tho' finely
written I could have wished had remained in Manuscript to be commu-
nicated only to a few people. When you read the work you will see my
reasons...", *The Correspondence of Adam Smith*, edited by E.C. Mossner
and I.S. Ross, Clarendon Press (Oxford, 1987).

immense machine, whose regular and harmonious movements produce a thousand agreeable effects";[148]

"In every part of the universe we observe means adjusted with the nicest artifice to the ends which they are intended to produce ... The wheels of the watch are all admirably adjusted to the end for which it was made, the pointing of the hour ... yet we never ascribe any such desire or intention to them, but to the watchmaker, and we know that they are put into motion by a spring, which intends the effect it ... produces as little as they do";[149]

"Every individual is led by an invisible hand to promote an end which was no part of his intention";[150]

"The rich ... consume little more than the poor ... They are led by an invisible hand to make nearly the same distribution of the necessaries of life ... The same principle, the same love of system, the same regard to beauty and order, of art and contrivance ... the great system of government, and the wheels of the political machine seem to move with more harmony and ease ...".[151]

"This universal benevolence ... all the inhabitants of the universe, the meanest as well as the greatest, are under the immediate care and protection of that great, benevolent and all-wise Being, who directs all the movements of nature; and who is determined, by his own unalterable perfections, to maintain in it, at all times, the greatest possible quantity of happiness. To this universal benevolence, on the contrary, the very suspicion of a fatherless world, must be the most melancholy of all reflections";[152]

---

[148] Smith, *Theory of Moral Sentiments*, op. cit., VII, iii, chap. 1,2, p. 316.
[149] *Theory*, II, ii, chap. 3, & 5, p. 87.
[150] Smith, *Wealth of Nations*, V, chap. 2, p. 456.
[151] Smith, *Theory*, IV, i, 10-11, pp. 184-5
[152] *Theory*, VI, ii, chap. 3, *Of universal Benevolence*, p. 235.

"The idea of that divine Being, whose benevolence and wisdom have, from all eternity, contrived and conducted the immense machine of the universe, so as at all times to produce the greatest possible quantity of happiness, is certainly of all the objects of human contemplation by far the most sublime ... the administration of the great system of the universe, however, the care of universal happiness of all rational and sensible beings, is the business of God and not of man";[153]

"the great Director of the Universe, the great Conductor of the Universe";[154]

"The idea of an universal mind, of a God of all, who originally formed the whole, and who governs the whole by general laws, directed to the conservation and prosperity of the whole ... As soon as the universe was regarded as a complete machine, as a coherent system, governed by general laws, and directed to general ends, viz. its own preservation and prosperity ...";[155]

"Every part of nature, when attentively surveyed, equally demonstrates the providential care of its Author, and we may admire the wisdom and goodness of God even in the weakness and folly of man";[156]

"The happiness of mankind, as well as of all other rational creatures, seems to have been the original purpose intended by the author of nature, when he brought them into existence. No other end seems worthy of that supreme wisdom and divine benignity which we necessarily ascribe to him; and this opinion, which we are led to by the abstract consideration of his infinite perfections, is still more confirmed by the examination of the works of nature, which seem all intended to promote

---

[153] *Theory*, VI, ii, chap. 3, 5, pp. 235-6.
[154] *Theory*, VI, ii, chap. 3, 4, pp. 235-6.
[155] Smith, *History of Ancient Physics*, in *Essays on Philosophical Subjects, Works*, III, 106 ff., ibid., &, pp. 113.
[156] *Theory*, II, ii, chap. 3, & 2, pp. 105-106. (ok)

happiness, and to guard against misery. But by acting according to the dictates of our moral faculties, we necessarily pursue the most effectual means for promoting the happiness of mankind, and may therefore be said, in some sense, to co-operate with the Deity, and to advance as far as in our power the plan of Providence";[157]

"As soon as the Universe was regarded as a complete machine, as a coherent system, governed by general laws, viz. its own preservation and prosperity, and that of all the species that are in it";[158]

"The happiness of mankind, as well as of all other rational creatures, seems to have been the original purpose intended by the Author of Nature, when he brought them into existence ... the works of nature ... seem all intended to promote happiness, and to guard against misery";[159]

"The great conductor and physician of the universe, has ordered to such a man a disease, or the amputation of a limb, or the loss of a child ... The harshest prescriptions of the great Physician of nature ... are indispensably necessary to the health, to the prosperity and happiness of the universe, to the furtherance and advancement of the great plan of Jupiter. Had they not been so, the universe would never have produced them; its all-wise Architect and Conductor would never have suffered them to happen".[160]

Joseph Priestley (1733-1804), theologian and chemist, the discoverer of oxygen and of several gases or 'airs', the author of writings on electricity, in his *Letters to a Philosopical Unbeliever*,[161] that is an answer to Hume, explains that the subject of his speculation is

---

[157] *Theory*, III, 5, 7, p. 166.
[158] *History of Ancient Physics*, III, pp. 106 ff, ibid., & 9, p. 113.
[159] *Theory*, III, chap. 5, & 7, p. 166. (ok)
[160] *Theory*, VII, ii., I, & 38, p. 289.
[161] First published in 1774. See now J. Priestley, *Works* (London, 1817-32; repr., New York and London, 1983), vol. IV, Part I, *Containing An Examination of the Principal Objections to the Doctrines of Natural Religion, and especially those contained in the Writings of Mr Hume.*

"whether the world we inhabit ... had an intelligent and benevolent author, or no proper author at all. Whether our conduct be inspected, and we are under a righteous government, or under no government at all".

Priestley's rational theology is deistic and opposing atheism. He maintains that "an atheist cannot have that sense of *personal dignity* and *importance* that a theist has". By contrast,

"the man who considers himself as a link in an immensely connected chain of being ... who considers as standing in the nearest and most desirable relation to a Being of infinite power, wisdom and goodness; a Being who gives unremitted attention to him, who plans for him, and conducts him through this life ... a man who really believes this ... must be another kind of being than an atheist".[162]

Then Priestley adds, having recourse to Ciceronian argumentations, that are transcribed almost word for word: "... wherever there is a fitness or correspondence of one thing to another, there must have been a cause capable of comprehending, and of designing that fitness ... that there is such a contrivance in the structure of a man's body, and especially something so wonderful in the faculties of his mind ... cannot be denied ... the human species must have had a designing cause ... all the species of brute animals ... all the visible universe ... must have had a cause or author, possessed of ... infinite power and intelligence ... the world must have had a designing cause, distinct from, and superior to itself ... This necessary cause we call God".[163]

Again, he insists on the "primary, intelligent cause ... an uncaused, intelligent Being ... Something must have existed from all eternity ...".[164] In Letter V, on 'The Evidence for the General Benevolence of the Deity', he mentions the "innumerable marks of design through the whole system of nature", which prove that "the author of it is intelligent, and, consequently, had some end in view

---

[162] Priestley, ibid., Preface, p. 317-320.
[163] Ibid., Letter II, pp. 331-3.
[164] Ibid. letter III, p. 334.

in what he did". At the same time, "means and ends are perpetually occurring to our observation ... we hardly see the hand of man without perceiving marks of design".[165] In Letter VI, on the 'Arguments for the Infinite Benevolence of the Deity', Priestley insists that "God's benevolence is infinite". Even death makes room "for a succession of creatures of each species". Even pain, that is "of real use with respect to true happiness", if considered "with respect to its future necessary effects ... the impressions of pain remaining in the mind ... contribute most of all to the future enjoyment of life". Furthermore,

> "the unlimited benevolence of the Author of Nature is not affected by the partial evils to which we are subject ... the whole plan of nature is admirable, chiefly on account of its being a system of wonderfully general and simple laws, so that ... the greatest good is produced with the least possible evil".[166]

Last, in Letter IX, the devastating attack upon the philosophy of Hume takes place. Priestley accuses him of promoting the cause of atheism. In the section on the 'Criticism of the Dialogues', he writes that "Philo speaks the sentiments of the writer", and advances "nothing but common-place objections against the belief of a God". Though the debate "closes in favour of the theist, the victory is clearly on the side of the atheist".[167]

---

[165] Ibid., pp. 344-6.

[166] Ibid., pp. 351-55, passim.

[167] Ibid., pp. 367 ff. In his sweeping attack against Hume, Priestley adds that "few men ever wrote with more perspicuity, the arrangement of his thoughts being natural, and his illustrations peculiarly happy; yet I can hardly think that we are indebted to him for the least real advance in the knowledge of the human mind". His object was "literary reputation, not the pursuit of truth, or the advancement of virtue and happiness; and it was much more easy to make a figure by disturbing the systems of others, than by erecting any of his own ... In many of his *Essays* ... Mr Hume seems to have had nothing in view but to amuse his readers ... proposing doubts to received hypotheses, leaving them without any solution, and altogether unconcerned about it". Furthermore he "never gave himself the trouble of reading Hartley's *Observations on Man* and the association of ideas ... so that to a person acquainted with this theory of the human mind, Hume's *Essays* appear the merest trifling", ibid., pp. 367-8.

One could conclude, again, that we don't need to add any comment to the texts, which speak for themselves.[168]

Nevertheless we must observe that Priestley, although he was a natural scientist, does not seem to move beyond Cicero in cosmology, on the general idea of the universe, on optimism, etc. And yet David Hartley, the founder of the associationist school of psychology, had published his *Observations on Man* in 1749, Condillac his *Traité des Sensations* in 1754, while in Germany the dogmatic philosophy of Christian Wolff was still prevailing, until the arrival of Kant's critical philosophy. The philosopher of Königsberg published in 1784 his *Answer to the Question: What is Enlightenment?*, and in 1793 his *Religion within the limits of Reason alone.*

With these authors, and with Kant in particular, we perceive that a different philosophical discourse has now taken the stage in European thought, and Ciceronianism begins to appear to be superseded.

Still Scotland produced some authors of note, Henry Home Lord Kames (1696-1782), and James Burnett Lord Monboddo (1714-1799).

Lord Kames, a prolific writer, was the author of the *Essays on the Principles of Morality and Natural Religion* (1751), of the *Elements of Criticism* (1762), in two volumes, of the *Sketches of the History of Man* (1774), his major work, in four volumes, and of other works, in which he debated subjects typical of the Enlightenment.

We will judge his *Sketches* with the words of James Boswell, who observed that the author "has a prodigious quantity of quotation,

---

[168] See nevertheless J.C. Stewart-Robertson, 'Cicero among the Shadows: Scottish Prelections of Virtue and Duty', in *Rivista Critica di Storia della Filosofia*, XXXVIII, 1983, pp. 25-49. Having consulted the sets of notes in the Scottish university libraries, i.e. in Aberdeen, St Andrews, Glasgow, Edinburgh, by the prelectors, J. Pringle, Th. Reid, Wm Cleghorn, D. Fordyce, A. Gerard, J. Mylne, Wm Law, J. Craigie, and Hutcheson's *Short Introduction to Moral Philosophy*, Stewart-Robertson comes to the conclusion that "the response of the Scottish prelectors towards the ancient moralist (Cicero) was clearly warm and at times even credulous". The *duties* as in the *De Officiis* "appeared neither dull nor disturbing to the minds and habits of the Scottish prelectors and, one presumes, of their students" (ibid., p. 48).

Therefore, the prelectors did not seem to be interested at all in Ciceronian cosmology.

and there seems to be little of what he gives as his own that is just, or that has not been better said by others."[169]

He collects a huge amount of materials, but no substantial line of narration appears in the four volumes. One can find all these materials, for example, in Adam Ferguson, where they are debated with a quite different mastery. Kames describes human history as having four distinct stages, thus contributing to studies on anthropology and sociology. But all these were topics current in literary debate. He does not need to draw many citations from Cicero, although the presence of the great Roman was constantly felt. And Kames writes, echoing Ciceronian concepts and wording, that

> "the Deity is the primary cause of all things. In his infinite mind he formed the great plan of government, which is carried on by laws fixed and immutable. These laws produce a regular train of causes and effects ... This universe is a vast machine, winded up and set a-going: the several springs and wheels operate unerringly upon one another. The hand advanceth and the clock strikes, precisely as the artist had determined ... In this plan, man bears his part, and fulfils certain ends for which he was designed".[170]

Lord Monboddo, in his *Ancient Metaphysics: or, the Science of Universals. With an Appendix containing an Examination of the Principles of Sir Isaac Newton's Philosophy*,[171] moves from the concept that everything has been said by the ancients, to which not much needs to be added. This is the arrière-pensée that leads him in his research, and is obviously absurd. The book is full of problems, but not equally full of responses to the problems its author raises.

For example, Monboddo writes that

> "the antient system had this advantage over the Newtonian, that it is more universal; so universal,that it is

---

[169] Boswell, *Correspondence*, III, 43. And James Beattie added that "a man who reads thirty years, with a view to collect facts in support of two or three whimsical theories, may no doubt collect a great number of facts, and make a very large book", cited by I.S. Ross, *Lord Kames and the Scotland of his day*, OUP (Oxford, 1972).

[170] Kames, *Essays*, the third ed. (Edinburgh, 1779), Essay III, p. 192.

[171] (Edinburgh, 1779-99).

well entitled to the name of philosophy, and the first philosophy, as it explains the principle of all motion, and all production in the heavens and in the earth".[172]

Then, moving from Aristotle to Cicero, he has recourse to Ciceronian concepts and terminology, and adds that

"man ... is set as a spectator in this great theater of the universe, where he is to admire the wonderful art and contrivance by which they are put together, so as to form one piece of amazing uniformity and regularity, as well as variety; to recognise his own and other minds, and by degrees to rise to the contemplation of that Supreme Mind, whose infinite goodness, wisdom, and power, have produced the wonderful scene presented to him".[173]

And again: the question is

"Whether the celestial bodies are moved by the immediate operation, and constant agency of Mind, according to the opinion of the antient philosophers? Or Whether, as the Newtonians maintain, the Heavens are a machine, which God Almighty had indeed contrived, framed, and set a-going, but which goes on of itself, without his interposition, or that of any other Mind, by the operation of causes merely material and mechanical"?[174]

The response, obviously, is that Newton was "a man of Science but no Philosopher",[175] and "did not know the Philosophy of Motion nor its cause".[176]

Nevertheless Monboddo, with his *The Origin and Progress of Language*,[177] was involved in the development of the theory of evolution, and is remembered nowadays as the founder of comparative historical linguistics.

---

[172] *Ancient Metaphysics*, Introd., p. III.

[173] Ibid., vol. I, book I, chap. 4, p. 102.

[174] Ibid., Appendix- 'Dissertation on the Principles of the Newtonian Philosophy', chap. 1, p. 498.

[175] On the same subject see above, note 134.

[176] Ibid., vol. sixth, chap. III, p. 27.

[177] In six vols. (Edinburgh, 1779-1799).

Late in the eighteenth, and early in the nineteenth century, the ageing Adam Ferguson set to jot down his manuscript essays which, in the cultural history of his years "represent a late chapter in the history of Stoicism, and an attempt at giving an answer to contemporary philosophical debate ... essentially on the basis of ancient philosophy. This still had its say in the eighteenth century ... Ferguson is a great explainer of historical change, and Stoicism was the philosophy of moral duty and also of political commitment, as anyone who is acquainted with Cicero's works, for example, knows well".[178]

The most relevant characteristic of the Enlightenment was optimism, with the faith in the destiny of man, in his fitting and good place in the system of the universe. This is advanced by Ferguson with plenty of arguments and novel concepts. To do so, he has recourse to Stoicism and to Cicero, whose philosophy he revived. And the premise of his reasoning, in these late essays is, as was usual in cosmology of his century, in *design* and *final causes*. Man, nevertheless, is not only part of design. By contrast, *this design is for him*. Hence "he must be conscious of his part, in his path 'to the Great scene of the Universe' (MS II, f. 7), because 'the Maker seems to have admitted him as a kind of Partner in his Work' (MS XVII, II, f. 18). Hence the Shaftesburian sections, where Ferguson's *new humanism* fully reveals itself. Hence the recurring references to the earth, 'on which Millions of Men are yearly Born'; to the 'Planetary System', which appears as a 'mere Scennery for the Reception and accommodation of Intelligent Being' (MS V, f 51; to the 'Myriads of human beings that every moment Nature sends into Life' (MS XIX, f. 3). 'Material Creation', therefore, is prepared only for the 'Reception of man' (MS VIII, f. 7), who is at the centre of this 'terrestrial Scene', which is 'the Lot of Man' (MS VIII, f.

---

[178]See V. Merolle, 'Adam Ferguson from Stoicism to the Philosophy of Enlightenment', Introd. essay, to Ferguson, *The Manuscripts*, op. cit., p. xi. The earliest of these essays (nos. II and XIII) are written on paper watermarked 1799, but the large majority of them are written on paper watermarked 1806 and 1808, and Appendix A in part on paper watermarked 1810.

In his *Principles of Moral and Political Science*, 2 vols., (Edinburgh, 1792), Ferguson writes: "The Author, in some of the statements which follow, may be thought to be partial to the Stoic philosophy; but is not conscious of having warped the truth to suit with any system whatever", ibid., Introd., p. 7.

1). Consequently, the design of nature is to furnish Materials for the exercise of the powers of his mind' (MS XXVII, I, ff. 18-21). Finally, the 'Universe of Body' is formed for the sake of mind alone, and to us appears to have 'no Other End or Purpose' (Ms VIII, f. 5)".[179]

In conclusion, one can say that these essays represent much more than the concluding step in the intellectual itinerary of their author. They represent a last, full contribution to the Enlightenment, to its cosmopolitan ideals, which were peculiar characteristics of Stoicism and of Roman tradition.[180] They were about to be superseded by Romanticism and the idea of *Nation*, in the German sense. Then one can say that finally Cicero's philosophy was superseded. Supposing, obviously, that twenty centuries of human thought can be superseded by new discoveries, which, in any case, leave humankind a 'prisoner' in this terrestrial globe.

Concluding remarks

This essay, primarily intended as a defence of Cicero against Theodor Mommsen, has developed into three distinct essays, although connected in a general reasoning.

The first part is a defence of Cicero against Mommsen. Nevertheless, to demonstrate that the German historian was not a real liberal, a short history of the principal ideas of liberalism has been sketched. This can be considered as an essay within the essay.

Thirdly, as regards cosmology, both Newtonianism and eighteenth-century philosophy are Ciceronianism, as elaborated by successive philosophers, but remaining substantially the same in its nucleus.

Ciceronianism is in fact superseded, supposing that it is, as we have said above, only by Kant and Hegel and by the physics of Einstein, which supplements the Ciceronian and Newtonian force

---

[179] I am transcribing from *The Manuscripts*, op. cit., Introd. essay, p. XX.

[180] A reference to the classical A. O. Lovejoy's *The Great Chain of Being* (1936) seems here necessary to conclude our reasoning. Lovejoy's work is essential to understand eighteenth-century optimism, the concept of the best of the possible worlds, etc., but its author does not seem to emphasize the role of Cicero's philosophy in these concepts. See in particular chap. VI, 'The Chain of Being in eighteenth-century thought, and man's place and role in nature'.

of gravity with its concept of curved lines in space, and with the theory of relativity. But the force of gravity remains as the fundamental concept in terrestrial physics, and cannot be replaced by new discoveries in this field.

These points, we believe, need to be emphasized, because they are mostly neglected by historians. It is in fact extremely difficult to grasp, all at once, twenty five centuries of human science, even for the most distinguished scholars. The only way of dealing with these problems has been, in the present essay, that of transcribing entire sections from the authors, who, in this way, speak themselves, instead of speaking through the words of interpreters, who do not always emphasize what is really meaningful in them.

Our commitment to these problems is owed to the charm that Cicero exerts upon us: as a wonderful writer, the finest mankind ever produced, along with Plato; as a precursor of Christianity; and, no less, as a defender of the republican liberties against tyranny. And Caesar was a tyrant. One can well try to justify his deeds as an *historical necessity*, as Hegel did. One can well believe that *Monarchy is the best*, as Drumann did.[181]

Nevertheless we live in an historical epoch that represents the triumph of the ideas of liberty, and that is rapidly getting rid of the last tyrants, who leave no heirs.

As our readers can see, ours is an essay on the history of ideas, which is our peculiar research field. And a passion for philosophical and political ideas has led us in our research, in our 'bill of indictment' against Mommsen, in our eulogy of ideas of liberty, during this happy period of our scholarly life, that we have spent with Cicero.

It remains to express the feeling of disappointment that we suffer when walking in Via dei Fori Imperiali, in this Rome of ours. This majestic road was opened in 1924-32 by our 'great Dictator', as a monument to the achievements of the 'regime'. And, obviously, the Dictator filled it with the statues of the emperors.

What seems unbearable is that, almost in front of the Curia, of which Cicero was the *princeps*, there is a statue to 'Iulio Caesari Dictatori Perpetuo', or to Julius Caesar the tyrant, as we shall call him.

---

[181] On Drumann see above, no. 6 & ff.

This statue should be removed. It represents in fact an affront to the believers in the ideas of liberty, of advancement of learning, of civilization.[182] And a statue to Cicero, the champion of Republican ideals, should be finally raised in its place.

---

[182]Let me cite here again from Cicero: "In primisque hominis est propria veri inquisitio atque investigatio", "above all, the search after truth and its eager pursuit are peculiar to man", *De Officiis*, I, iv, 13.

Chapter II: Bibliographical essay

To speak about Cicero and Ciceronianism in modern times, from the Enlightenment onwards, one needs to begin with Conyers Middleton's *Life of Cicero* (1741) as the standard work that, in Ciceronian bibliography, preceded both Drumann and Mommsen, was soon translated into Italian and French, and went through several editions.

The book "relies heavily upon paraphrases and long extracts of Cicero's own writings, taking most of what Cicero says about his own achievements at face value", and providing "a minimal narrative framework", as Matthew Fox writes.[1]

It certainly "has laid before us a very dry list of facts",[2] and cannot be considered as a critical biography. Notwithstanding its limitations, it was successful in establishing a reputation for Cicero. In fact, we can observe, it emphasized the most relevant points in his biography and thought, and this very character ensured its success. Many books are simply works of documentation, and their usefulness is in their giving the information one needs. In the eighteenth century people needed to know more about Cicero, whose influence was deeply felt in the world of learning, and Middleton came out with his biography at the right moment, submitting to the readers exactly the contents they wanted to know.

Nevertheless in Great Britain the debate about the figure and work of Cicero was impassioned. Addison Ward, in his 'The Tory View of the Roman History', describes "a violent wave of protest against the idealization of Cicero and his party which arose toward the middle of the eighteenth century in reaction to Conyers Mid-

---

[1] See M. Fox, 'Cicero during the Enlightenment', in C. Steel ed., *The Cambridge Companion to Cicero* (Cambridge UP, 2013), pp. 318-337, ibid. p. 332.

[2] So the contemporary 'Oxford scholar', cited by M. Fox, ibid. But Sir Leslie Stephen considers Middleton as "the true precursor of Gibbon", thus recognizing the merits of his *Life*: see L. Stephen, *History of English Thought in the Eighteenth Century*, Peter Smith (New York, 1949), I, p. 270.

dleton's eulogistic *Life of Cicero*".[3] In particular, the attack culminated in Nathaniel Hooke's monumental *Roman History, from the Building of Rome to the Ruin of the Commonwealth* (1738-1771), which presented Julius Caesar as the hero, and the defenders of the Republic as "narrow-minded spokesmen of a corrupt oligarchy". In Hooke's history "the Senate is the oppressor, and popular *encroachments* ... are steps toward, not away from, Liberty. With this reversal, the reputation of Cicero receives the *coup de grâce*."[4] One had to wait until Adam Ferguson's *History of the Progress and Termination of the Roman Republic*[5] to find that Caesar's reign was a period of lawless tyranny, and his death a lesson to tyrants.

In the middle of the next century, in 1854-56, Mommsen's *Römische Geschichte* was published. It had an enormous influence and is still widely read, because of its author's literary achievements and the fascinating vigour of his style. As Karl Christ has written,

> "it has remained until today as the classical narrative of the history of the Roman Republic in German, a literary and scientific masterpiece, that still enjoys the greatest authority".[6]

We have mentioned the answer that Gaston Boissier gave to Mommsen, in his *Cicéron et ses Amis* (1865), but to the list of answers to Mommsen we need to add J.L. Strachan-Davidson, *Cicero and the Fall of the Roman Republic*.[7]

In the Preface to his book Strachan-Davidson maintains that "in writing Roman history it is impossible to escape from the influence

---

[3] *Studies in English Literature*, 4.3 (1964), pp. 413-56, ibid., p. 414.

[4] Ward, ibid., pp. 448-9.

[5] Printed for Strahan ' Cadell, in the Strand, and W. Creech in Edinburgh, 1783. "*The Roman Republic*, if considered as a piece with the *Essay* and with the *Principles*, could be defined as "a long paean to republican virtue, a sustained example of what civic virtue really means", see V. Merolle, Introductory Essay to *The Manuscripts of Adam Ferguson*, Pickering & Chatto (London, 2006), p. xxi.

[6] "Sie blieb bis zum heutigen Tag die klassische Darstellung der Geschichte der Römischen Republik in deutscher Sprache, ein literarisches wie wissenschaftliches Meisterwerk, das noch immer höchste Autorität genießt", see K. Christ, 'Theodor Mommsen und die Römische Geschichte', in Th. Mommsen, *Römische Geschichte*, Vollständige Ausgabe in acht Bänden, Deutscher Taschenbuch Verlag, Band 8, S. 7.

[7] New York and London, 1902; repr. in Forgotten Books, 2012.

of the genius of Mommsen. Sometimes by suggestion, sometimes by repulsion, his presence is always felt". He adds that the purpose of his volume is "to tell the story of Cicero's life, and at the same time to set forth from his writings a presentation of the concluding age of the Roman Republic and to record *the disastrous but not inglorious failure of the last Free State of the ancient world*".[8]

These words are an intimation of the character of Anglo-Saxon historiography, i.e. the defence of the Roman Republic against incoming tyranny. This is the reason why the influence of Cicero's ideas was largely felt by the American founding fathers.

John Adams, second President of the United States (1797-1801), wrote that "all the ages of the world have not produced a greater statesman and philosopher united than Cicero".[9] Thomas Jefferson read *De Senectute* every year, Josiah Quincy owned three editions of Cicero, "testifying to his devotion to Cicero as a defender of liberty". Jonathan Mayhew avowed that the chief sources of his ideas of civil liberty were Plato, Demosthenes and Cicero.

> "These high estimates of the Roman were general among John Adam's eighteenth-century contemporaries both in England and in the America that was part of the English intellectual world. Cicero spoke with such convincing force to that world because in many ways he belonged to it".[10]

Both English and American historians are on this line, and virtually no one of them has ever criticized Cicero, who has constantly been felt as the defender of Republican ideals and of ideas of liberty.

We need to mention, in particular, Henry Hart Milman who, in *The London Quarterly Review*, 1836, promptly defended Cicero against Drumann, and Moses Stephen Slaughter who, a century later, in 1921-22, in *The Classical Journal* descended into the arena against Mommsen, with vigorous words that, borrowing from him, we can call *enraged*.[11]

---

[8] Italics mine.

[9] 'A Defence of the Constitution of the United States of America', in *The Works of John Adams*, vol. IV.

[10] H. J. Haskell, *This was Cicero. Modern Politics in a Roman Toga*, Secker & Warburg (London, 1941), p. 8.

[11] Both these essays are reprinted here, in Appendix A.

Another milestone in Ciceronian literature was Thäddäus Zielin-ski's *Cicero im Wandel der Jahrhunderte.*[12] Zielinski, a Polish Ukrainian, was professor in St Petersburg and Warsaw. Born in 1859 near Uman, in Kiev Governatorate, he died in 1944 in Schorndorf, Upper Bavaria.[13] His book, first published in 1897, was a pioneering work, and as a comprehensive view of the subject it still has its own validity, as shown by successive editions and reprints. It has been superseded only in part by more recent research.

The author insists in particular on eighteenth-century deism as finding its origin in Cicero's works, writing with Voltaire that "das älteste Denkmal des Deismus waren Cicero's Bücher *De Natura Deorum* und ihr Anhang *De Divinatione*", "the most elevated monuments of Deism were Cicero's books *On the Nature of the Gods* and its appendix *On Divination*". He adds that "seine *Tusculanen* sowie sein Buch *Von dem Wesen der Götter* die beiden schönsten Werke sind, die menschliche Weisheit jemals verfaßt hat", "his *Tusculans* not less than *On the Nature of the Gods* are the most beautiful works that human wisdom has ever produced".

Again with Voltaire, he repeats that Cicero's works "will safeguard the world from despotism", "werden die Welt vor dem Despotismus schützen".

And, last:

"Wenn die monarchische Gewalt sich gefestigt haben wird, darf man wohl erwarten, daß sich unter diesen Tyrannen auch einzelne gute Herrscher finden lassen ... wenn sie (die Völker) deine Schriften lesen, werden sie

---

[12]I am using the Sechste Auflage (Stuttgart, 1973), Nachdruck der 3. Auflage (Leipzig, 1912). Along with Zielinski's book should be cited Remigio Sabbadini, *Storia del Ciceronianismo* (Torino, 1885).

[13]His son-in-law was Vladimir Beneshevich (1874-1938), an outstanding scholar in the field of Byzantine history, whose book on John Scholasticus appeared in a German edition in Munich in 1937. In October an article on the *Izvestia* portrayed this as a betrayal. On 17 January 1938 Beneshevich, together with his twin sons and his brother, was executed by a NKVD firing squad on charges of spying for Germany.

On 20 August 1958, twenty years after his death, he was exonerated of the charges of treason by a Military Tribunal, and on 19 Dec. of the same year was rehabilitated by the Russian Academy of Sciences. Beneshevich is one of the Eastern Orthodox Church's New Martyrs.

tugendhaft werden", "if monarchic power will consolidate itself, then people can expect that even under these tyrants will it be possible to find single good rulers ... if reading your writings, (nations) will become more virtuous."

Zielinski then cites from a letter by Friedrich der Große to Voltaire:

"Niemals hat es auf der Welt einen zweiten Cicero gegeben ... ich liebe Cicero unendlich ... ich finde in seinen *Tusculanen* viele Gefühle, die mit den meinigen verwandt sind ... das Buch *Über die Pflichten* ist das beste Werk auf dem Gebiete der ethischen Philosophie, das jemals geschrieben worden ist oder geschrieben werden wird", "never has there been in the world a second Cicero ... I love Cicero immensely ... I find in his *Tusculans* many feelings that are similar to my own ... the book *On Duties* is the best work in the field of ethical philosophy, that has ever been written, or will ever be written".[14]

Obviously, Zielinski did not need to bring forward any open polemic with Mommsen. His book is in fact in itself an explicit criticism of the German historian, especially when emphasizing the judgments of Voltaire and Friedrich der Große on Cicero.

Twentieth century and beyond

The Ciceronian bibliography in the twentieth and in the beginning of this twenty-first century is extremely rich. We can say in fact that we are now living a Cicero-revival, with the progressive achievement of ideas of liberty and advancement of progress in this western part of the world.

A bibliography that aims at becoming exhaustive can be found on the internet site www.Tulliana.eu of the S.I.A.C., or *Société Internationale des Amis de Cicéron*, c/o Ermanno Malaspina, of the University of Turin. This site is under construction, and will need

---

[14]Zielinski, ibid., 247-8.

years to reach completion.[15] Nevertheless it responds to an original idea, in line with both the times and modern technology.

In *The Cambridge Companion to Cicero*,[16] edited by Catherine Steel, that sums up the present state of Ciceronian studies, in the chapter on 'Twentieth/twenty-first-century Cicero(s)' Lynn S. Fotheringham lists twenty-four Cicero biographies, all in English, by British and American authors, excepting two, by German authors, but translated into English. All of them variously contribute to the knowledge of the Roman statesman, philosopher, orator.

Some of these biographies are the work of non-academic authors, nevertheless they have merits. For example, Henry Joseph Haskell[17] was an American journalist and, as we learn from the title of his book, he aimed at understanding 'modern politics in a Roman toga', drawing parallels with the contemporary world and with American politics in particular. He writes that in a previous book he was impressed

> "with the reappearance in modern America of many of the economic problems of ancient Rome. I was impressed as well with the persisting character of the social pressures that have compelled governments, ancient and modern, to adopt very similar measures in dealing with these problems".[18]

In connection with Haskell's book, one can think of Hume's words:

> "It is universally acknowledged that there is a great uniformity among the actions of men, in all nations and ages ... Would you know the sentiments, inclinations, and course of life of the Greeks and Romans? Study well the temper and actions of the French and English ... mankind are so much the same, in all times and places, that history informs us of nothing new or strange in this particular".[19]

---

[15] Try in particular under http://www.Tulliana.eu/ephemerides/altro/bibliografia.
[16] Cited above, n. 1.
[17] See above, n. 10.
[18] Haskell, preface, p. III.
[19] Hume, *Enquiries*, ed. Selby-Bigge (OUP, 1975), 'Human Understanding', Sect. VIII, 'Of Liberty and Necessity', Part I, & 65.

Many pages of Haskell's book are pure journalism, and we will not follow its author in insisting on them. As a political journalist, he has his own eye cast on British and American politics, therefore he often gives useful teaching to the readers who are interested in social and political ideas. For example, when mentioning Cicero's concept of law, he comments: "It was this philosophy that infused the American Declaration of Independence as formulated by Thomas Jefferson".[20] When speaking about Catiline, he tries to defend Catiline's reasons in defense of the wretched who could not pay their debts, observing that his could have been just "a reasonable and legitimate protest", and "it may be assumed he thought of himself as leading the cause of free men".[21]

These points recall to the attention of the reader what we can call 'the rhetoric of the winners', of which we have numerous examples in history. For example, how many people know what the social democratic Kerensky government in 1917-Russia was, and what it achieved? The victory of Lenin and Stalin handed Russia over for generations to backwardness, preventing the natural energies of that society from operating in favour of progress and of the advancement of civilization. And yet, in southern Europe in particular, for half a century, after the war to the fall of the Soviet Union, we were victims of the 'rhetoric of the winners', which heavily conditioned the life of many of us.

Another book owing to a non-academic historian is *The tremulous hero. The age and Life of Cicero. Orator, Advocate, Thinker and Statesman*[22] by A.F. Witley, pseudonym of Sándor Forbát, Hungarian doctor and poet: a literary masterpiece, indeed. Forbát is in fact an astonishing writer, and his art immediately surprises the reader, appearing to its full extent in the preface, in his initial eulogy of Rome. Like most non-professional authors, he has his eye

---

[20] Haskell, op. cit., p. 67.

[21] According to Haskell "there are other bits of evidence that the movement for the revision of debt legislation was regarded by those who took part in it as merely a reasonable and legitimate protest ... It may be assumed that Catiline thought of himself as leading the cause of free men ... Catiline and his followers began to discover that they were being manoeuvred into the position of rebels and so took what steps they could to defend themselves ...", Haskell, op. cit., p. 183.

[22] Pallas Publishing (London, 1949).

cast on the present, and this attitude is particularly helpful in explaining what ancient Rome was. He observes in fact that the spirit of the age was that of today, where "it will be easy to find in Caesar, Pompey, Crassus, Catiline and Cicero of those days the Hitlers, Mussolinis, Goebbels, Görings and Cianos of our own times".[23]

This observation, along with many others interspersed in the pages of the book, gives the clue to understand his work, which has its weak points if considered as a contribution to history as a science, but helps the reader realize what human society was, and is. This is the most relevant character of many works by non-academic historians, and Witley's in particular.

The last section of the book is substantially a paean to Cicero, when its author writes that

> "Cicero appears as compared with Caesar, as decidedly the more modern, the more intellectual, and the more advanced of the two ... Cicero's achievements will ever appeal more irresistibly to the modern mind. He was already a citizen of a modern state, anxious to promote and secure the freedom of the individual ... as the champion of liberty and progress ... Caesar only laid the foundation of the Roman Empire, which has since passed away. But Cicero has been the interpreter of the Graeco-Roman culture for thousands of years and, maybe, for all time. Passionately convinced of the dignity of the human soul, he speaks and thinks in the language of every civilized man ... *The world would have been poorer but might have done without Caesar, but if Cicero had not lived, the loss of mankind would have been much greater*".[24]

This is substantially the perception of Cicero, both the man and politician, in Anglo-Saxon historiography, apart from very few perplexities, which do not become open criticism.[25]

---

[23] Ibid., p. 18.
[24] Ibid., pp. 306-307. Italics mine.
[25] We have in mind, in particular, Lloyd Thompson and Sir Ronald Syme.

German biographies

As for German historiography, apart from the perplexities of Walter Rüegg,[26] only with Manfred Fuhrmann, with whose work we have already dealt,[27] does the substantial recantation of the world of learning in that country begin.

Fuhrmann tries to explain Mommsen's disparaging judgment of Cicero's works and person with the tradition of the Rhetorikverachtung, or contempt for rhetoric, proper to German tradition,[28] going back to Kant and Hegel, and variously appearing in the subsequent bibliography. Nevertheless his defence, or attempt at finding a justification, is an embarrassed one, and far from meeting the target, because he ignores entirely the political implications of Mommsen's judgment. After all, Mommsen knew well, extremely well, nineteenth-century European society, what he was writing, and what he aimed at.[29]

---

[26]See above, 'Mommsen and Cicero', n. 19.

[27]On Fuhrmann see above, 'Mommsen and Cicero', chap. 1, text corresp. with notes 19-24.

[28]The judgment of Paul Groebe, Drumann's editor, is curious and bizarre. He writes the following words: "Schon die einfache Tatsache, daß bisher niemand den Versuch gewagt hat, die Anklageschrift Punkt für Punkt zu widerlegen ... schon diese Tatsache allein läßt darauf schließen, daß die Vorwürfe berechtigt sind", "the fact itself, that until now no one has ventured to disprove the allegations point by point ... this fact itself brings us to the conclusion that the allegations are justified."

Groebe adds: "Man hat im Süden andere Vorstellungen von Recht und Unrecht, von Wahrheit und Unwahrheit, von Treue und Untreue als bei uns", "people in the south have different ideas of what is right and wrong, of truth and untruth, of loyalty and disloyalty than to us", see 'Vorrede zur zweiten Auflage', (Hannover, im August 1929), S. X.

This is a pretentious portrait. An answer to such words is out of the question.

[29]"Die bisherigen Betrachtungen sollten dartun, daß das negative Cicero-Bild à la Mommsen ... nicht unmittelbar seiner Person entsprungen sind, daß sie vielmehr durch seine Anwaltsberuf ... bedingt zu sein scheinen ... Die Attacken gegen die Redekunst ... nahmen im neunzehnten Jahrhundert eine bis dahin unerhörte Schärfe an ... in Deutschland", "the previous observations should demonstrate, that a negative portrait of Cicero, like that of Mommsen, did not rise directly from his person, but appears much more owing to his profession of advocate. The attacks against the art of rhetoric in Germany in the nineteenth century reached a sharpness until then never heard", Fuhrmann, 'Die Tradition', op. cit., p. 46.

Fuhrmann's book invites further reflection. It being written substantially for the German scholarly public makes the reader think how Anglo-American and German historiography go along two parallel roads, that will inevitably meet in the future, at some point of the progress of learning, we believe.

The book is substantially a biography, that should be more problematic (I am thinking in particular of Karl Rosenkranz's *Hegel's Leben*, and of E. C. Mossner's *The Life of David Hume* as the models of modern biographies). It has therefore many limitations, its author being more a philologist than an 'authentic' biographer. Nevertheless it has the merit of representing a turning point, of opening into the right road for the assessment of Cicero's person and philosophy in Germany. In fact it completely ignores Drumann, and almost completely ignores Mommsen, who is only mentioned in the last page, and only in order to keep its own distance from him. Therefore, it represents the definitive 'burial' of Mommsen's unfair judgment of Cicero, although not of him as a historian, as a philologist, as a writer.

Next, in *Cicero der Politiker*,[30] Christian Habicht also complains that "Cicero leidet noch heute, wie im 19. Jahrhundert, unter unsächlicher und hämischer Kritik", "Cicero still suffers nowadays, as in the nineteenth century, from an unobjective and malicious criticism",[31] owing to Drumann. "Das ist lange so geblieben... unter den Historikern als unter den Philologen, und am stärksten unter deutschen Historikern", "it has long lasted so ... under historians no less than philologists, and particularly among German historians", and "dank Mommsens Autorität wurde dieses negative Cicerobild nahern kanonisch", "owing to Mommsen's authority this negative portrait of Cicero became almost canonical".[32]

The aim of Habicht is that of demonstrating that Cicero "mit Leib und Seele Politiker war", "was with body and soul a politician", and this leads him to polemize with Christian Meier's contrasting opinion.[33] His polemic becomes most interesting when he mentions Andreas Alföldi's judgment, according to whom the killers of Caesar

---

[30] Beck publishers (München, 1990). The book was published simultaneously in English as *Cicero the Politician* (Baltimore, 1990).
[31] Ibid., p. 9.
[32] Ibid., p. 12.
[33] See Habicht, p. 7 and pp. 13-14.

were "diese feige Rotte", "a cowardly gang", while the assassination was "ein feiges Verbrechen in der Art der Mafia-Morde unserer Tage", "a cowardly crime similar to the Mafia-murders of our own days".[34] Unfortunately, Alfred Heuß, the celebrated Mommsen biographer, influenced by his demigod, Mommsen, we mean, also wrote that Cicero "war nicht der Mann, die Idee dieser Republik selbst zu verkörpern. Er wußte von ihr, aber er war sie nicht", "Cicero was not the man who could embody the idea of this Republic. He knew about it, but he was not it".[35]

It is a pity that both Alföldi and Heuß, along with a few others, could give such a judgment of Cicero. To understand Cicero, at least from the point of view of political ideas, one must be familiar with the history of political thought, that is substantially a history of philosophy, from the point of view of human society and the state, from Socrates and Plato to Marxism, via authors like Augustine, St. Thomas, Dante, Marsilius, Milton, Locke, the eighteenth century, -especially Montesquieu and Adam Smith-, the French authors of the nineteenth century, not forgetting German Romanticism and, finally, Marxism that, philosophically, is modern materialism.

Only if one has such a cultural background, can a scholar read, and truly understand, Cicero. Then he will reach the conclusion that Cicero was one of the noblest souls who ever lived and, along with Plato, the finest writer mankind has ever produced. If one does not have a proper perception of what politics is, no scholar can be able to understand him. And Mommsen, with a number of other scholars, outstanding philologists indeed, is extraneous to this world of ideas.

As for Christian Habicht, we can conclude with the words of the author, according to whom the book is "in erster Linie für Studenten der Geschichte und für Leser geschrieben die, ohne Fachleute zu sein, Interesse für den Gegenstand haben", "written first of all for students of history and readers who, without being specialists, are interested in the subject".[36]

---

[34] Habicht, 'Anmerkungen', p. 147, n. 30.
[35] See Habicht, *Epilog*, p. 119.
[36] Vorwort, p. 7.

These words testify to the new interest for Cicero in the German world of learning, that is finally celebrating its own liberation from Mommsen's tutelage, a tutelage that still haunts scholars as a ghost.

Stephanie Kurczyk's *Cicero und die Inszenierung der eigenen Vergangenheit*[37] is a literary study. Its focus is that of Cicero's *staging of his own past*, a subject recurrent in most recent literature.[38] It puts at the centre of analysis the autobiographical discourses and writings of the great orator, whose aim, the author maintains, is not 'Selbsterkenntnis', knowledge of himself, but *Selbstinszenierung* and *Selbstpräsentation*, staging and presentation of himself.[39]

Kurczyk's is a sophisticated study, and of particular interest is her examination of the allegations, recurring in the hostile historical literature, against Cicero's self boasting, whose meaning appears in a different light, when it is seen in a purely literary context.

Wilfried Stroh's *Cicero, Redner, Staatsmann, Philosoph*,[40] appearing in the series 'Wissen' of Beck Publishers, is an eloquent work, and it clearly responds to the desire of learned German society to know more about Cicero.

The thesis of the book is that Cicero was "ein politischer Platoniker", "a political Platonist", and in this way its basic concept moves beyond the article on *Marcus Tullius Cicero* of the *Realencyclopädie der Klassischen Altertumswissenschaft* of 1939.

Certainly the thesis of Cicero as a political Platonist is fully demonstrated by Stroh, nevertheless the concept of Stoic commitment to politics, to the defence of state and society against Epicureanism, should be emphasized, in his political ideas. If emphasizing this aspect Cicero will appear as quite modern, quite topical, fully responding to the needs of society today.

As for the *vexata quaestio* of Caesar's role in Roman history, that was history of the world then known, Stroh writes that, as a philologist, he will not try to solve the problem whether a monarchy of Caesarian inspiration would have been able to organize the Roman Empire. After all, Caesar and Augustus were "doch zunächst und

---

[37] *Cicero und die Inszenierung der eigenen Vergangenheit. Autographisches Schreiben in der späten Römischen Republik*, Böhlau Verlag (Köln Weimar Wien, 2006).

[38] See below, text corresp. with nn. 60-66.

[39] Kurczyk, ibid., pp. 15-16.

[40] Beck (München, 2010).

vor allem begnadete Machtmenschen, ohne tiefere Pläne globaler Beglückung", "first of all gifted power-seeking men, without a sound project of global happiness". Therefore, "honour to Cicero, who largely managed to lead a resistance against them", "es ehrt Cicero, dass er dem einen weithin Widerstand geleistet hat".[41]

Again, Stroh's book leads the reader to reflect upon the idea of historical necessity, as found in Hegel's philosophy, in Caesar's seizing power and restoring order in Rome. But Hegel's philosophy of history is *giustificatrice* non *giustiziera*,[42] *justificative* not *avenging* (in Schiller's words the 'Weltgeschichte' is 'Weltgericht', history herself is a court of justice, which renders justice to all the actors).

Substantially Hegel moves from the traditional metaphysics, especially in the *Phaenomenology*, that represents the course of events as unfolding from the Spirit, or God of traditional religion. Hence the Hegelian saying according to which philosophy is like the owl of Minerva, that begins flying only at sunset: i.e., it is reflection on what has already happened, not on what is happening now, in the present state of society.

Moving from Hegelian premises, everything finds a justification or, at least, an explanation. In this light Cicero is guilty of opposing the unfolding of the 'Spirit of the world', and needs to be defeated in his battle against historical necessity.

The philosophical movement opposing Hegelianism is philosophy of praxis, so well formulated by Karl Marx when he wrote, in the eleventh of his *Thesen über Feuerbach*, "die Philosophen haben die Welt nur verschieden *interpretiert*, es kömmt darauf an, sie zu *verändern*", "until now philosophers have only variously interpreted the world, the point is that we must change it".

This is no more than actual politics, and represents the point of view, or ideology, of a political party, not the reality of the whole of society. More philosophies of *praxis* (praxis meaning, in ancient Greek, doing, transaction, business) are constantly in action in society, each of them representing the interests of particular ranks or social classes. Therefore, it is difficult to consider true philosophy as a set of ideas or, rather, ideologies, that aim at changing the

---

[41] Epilog, p. 119.

[42] On this subject see B. Croce *Saggio sullo Hegel* (first ed., Bari 1906), passim.

world, instead of *understanding* it, instead of *explaining* the nature of historical events, without the pretence, often absurd, of judging their actors.

Shall we repeat that, paradoxically, Cicero's concept of the mixed constitution, considered as a purely political theory, is the most modern of all, even if doesn't have behind it two millennia of elaboration, unlike Hegel's philosophy? It does not represent the point of view of just one political party, but that of all the political actors, who all have their role to play on the stage of history. Similarly, it embodies the idea of the progress of society, the natural course of civilization of mankind, that cannot be thwarted by deviations and errors, so frequent in human history. Again, we think of Fascism, Nazism, Communism, which disfigured the European world in the twentieth century.

Therefore, let us repeat with Stroh: 'Ehre an Cicero', honour to Cicero, who fought to death for this fundamental, irreplaceable ideal, which should enlighten the rulers of human societies.

Stroh ends his book by trying to judge the errors of Cicero, but this lets the reader think, with Rousseau, that 'democracy is proper to a people of Gods'. After all, who can be sure that the 'errors' of Cicero were actually 'errors'? Certainly it was an error, on his part, to believe in the young Octavian, who affected deference and respect for him, calling him 'father'. But what else could the ageing Cicero have done, in such a tragic moment as the fall of the Roman Republic, apart from hope, vainly indeed, to direct this young rascal, who apparently was the most flexible, but who was the ablest and the worst of all? And should we condemn Cicero for having been deceived? His immense, unbounded passion for liberty should instead be praised.

To conclude with our reflections in connection with Stroh's book, human passions are difficult to restrain, particularly at some moments, in historical events. Are they so owing only to social and material reasons, as the Marxist historians would say? Or are they owing to several concurring factors, as the Liberal historians would say in their turn?

To give an answer to this question is beyond human understanding. Judging history is like judging our neighbour. We should refrain from doing so, because we should be conscious that our judgment is often warped by passions, prejudices, points of view, proper to any human being. But no one should condemn Cicero's love for liberty.

Klaus Bringmann opens his Cicero[43] by declaring that the book "ist nicht Fachleuten, sondern einem allgemeinen, interessierten Publikum gewidmet", is not for specialized readers, but for a general public, interested in Cicero.

With these initial words the author further tolls the bell for Mommsen, and emphasizes, against Alfred Heuß, that Cicero "am Ende die Idee der Republik verkörperte",[44] "Cicero personified to the end the idea of the Republic".

Any book invites reflection, and Bringmann's work, when narrating Cicero's desperation while in exile, makes the present author think of the tragedy of the political exiles under Fascism, Nazism, Communism.[45]

One needs to know well the history of the exiles, through documents, obviously, their feelings of dejection and humiliation, to understand Cicero's desperation when his house was devastated, as was his brother's house, when his wife Terentia had to take refuge with the little child in the temple of the Vestal Virgins, and the gang of Clodius also planned to seize her own property. "Thou shall leave each thing/ Beloved most dearly.../ Thou shalt prove/ How hard the passage, to descend and climb/ By other's stairs".[46] With these words Cacciaguida foretells to Dante his exile in the

---

[43]Primus Verlag (Darmstadt, 2010).

[44]Ibid., p. 294. The aim of the book, as the author explains it, "Ciceros Leben und Werk sind auf das Engste mit der Geschichte der späten Republik verflochten, diese Verflechtung zu analysieren und zur Anschauung zu bringen hat der Verfasser dieser Biographie sich zum Ziel gesetzt", "given that Cicero's life and work are intertwined in the closest way with the history of late Republic, the author's aim has been that of analyzing this connection" (ibid., p. 17).

[45]As for Italy, whose contemporary history was long the research field of the present author at the beginning of his scientific career, the exiles used to be pointed out, in their own nations, as traitors, with their families. Unfortunately also the public opinion of the countries where they had taken refuge, diverted by the propaganda of the regime, with difficulty could understand their tragedy. For example, in Great Britain public opinion credited Mussolini with representing a check to Communism, and the exiles were often seen with suspicion. Historical documents give ample evidence to this.

[46]Dante, *The Divine Comedy*, *Paradiso*, lines 57-60. Translation by H. F. Cary.

Paradiso. And words like these should lead any sound historian, any sound man, in his judgment of Cicero: guilty, if he was guilty of something, of being the greatest letter-writer mankind ever produced, of being *naturaliter* able to express in their fullness his own feelings of desperation to his wife, to the trusted friend Atticus, to his brother Quintus, in such a literary way that no one has ever been able to surpass.[47]

To conclude, Bringmann's biography is an outstanding one. The author knows the sources well and writes well. And the natural epilogue is a real panegyric of Cicero's figure and work.

An important aspect of the aim of the book and of its nature is found in the fact that the author calls Matthias Gelzer "der Verehrer Caesars", "Caesar's admirer", adding that in his *Cicero. Ein biographischer Versuch*, Gelzer puts Cicero "ganz im Schatten Caesars und erscheint er als an antiquierter, letzlich unpolitischer Politiker",[48] puts Cicero "wholly in the shadow of Caesar, and makes of him an old fashioned, substantially an unpolitical politician".

Arnd Morkel's *Marcus Tullius Cicero. Was wir heute noch von ihm lernen können*,[49] is a book structured around problems, i.e. its author presents problems and tries to give them an answer. From this point of view the book represents an original contribution to our knowledge of Cicero.

For example, as for Kant's and Hegel's judgments, Morkel defines them *Schubladendenken*, stereotyped thinking, and explains what eclectic philosophy is, i.e. that "Wahrheit an keine bestimmte Autorität und keine bestimmte Lehre gebunden ist", "truth is not bound to any particular authority or teaching"; or even, it is "die selbständige und unvoreingenommene Auswahl aus den Erkenntnissen anderer ... zur Annäherung an die Wahrheit oder auch zur

---

[47] We have to mention here, again, Jerôme Carcopino, whose *Les Secrets*, op. cit., was a sequel of spiteful judgments, from the first to the last page, such that reading the book is a hard task.

Carcopino joined the regime of Vichy, and in his capacity as a state administrator he applied the laws that excluded Jews and Freemasons from civil service. After the war he was jailed and brought to trial before the Supreme Court, but finally acquitted. As a historian he was certainly clever, even outstanding, but his book gives evidence of his warped way of reasoning, and proves what an ambiguous figure he was.

[48] Bringmann, p. 204 and 293.

[49] Königshausen & Neumann (Würzburg, 2012).

Gewinnung der Wahrheit", the "independent and unprejudiced selection from the knowledge of the others ... for the approach to truth or for its obtainment".[50]

Philosophers do not like the definition of eclectic, because it is synonymous with unoriginal thinking. Nevertheless, for the attainment of truth, some eclecticism appears necessary. Knowledge is a progressive conquest, and every thinker brings his own contribution to it. Philosophical systems often thwart the truth, due to the need of being consequent with the premises of the reasoning of their authors. But philosophical premises are often a postulate of reason, that, as such, cannot be proved. And, in any case, all of us, no one excluded, bring their own grain of sand to the edifice of truth.

No philosophy, therefore, represents absolute truth. Neither that of Hume, nor that of Kant or Hegel. Each system answers to the problems posed by a particular historical epoch, the 'great' systems are no more than the synthesis of the science of a whole epoch.

We can touch on here just some of the points made by Morkel. For example, the question, debated in the schools, following Seneca's *Suasoriae*, whether Cicero "was born too late and, if he would have been successful, had he been born in another epoch". Or the question, whether he was a good politician or not. Morkel cites the negative or restrictive judgments of Mommsen, who "not only hates Cicero, but despises him", of Sir R. Syme, M. Gelzer, Ch. Meier, W. Dalheim, who all maintain that Cicero was lacking any political genius.

To these authors we will observe that, from the laboratory of the historian, one can construct any type of bizarre judgment. Nevertheless Mommsen, so sure of himself in his contemptuous attitude towards Cicero, and along with him Syme, Gelzer, Meier, Dalheim, don't seem to understand what politics is, what society is, what historical change is. They lack any real perception of politics, as the science of society. A wider knowledge of history, of what society and human beings are, is gained through anthropology, and is necessary before setting out to judge Cicero. Without such a

---

[50]These definitions are by N. Hinske and M. Albrecht, cited by Morkel, ibid., p. 48.

knowledge scholars "should be advised to study in literary matters a becoming silence", to use Mommsen's words.[51]

After all, the question whether Cicero was or was not a good politician, is an idle one. Cicero was the author of those philosophical works whose concepts were deeply implanted in his person, in his heart, in a word, were personified by him. How can one conceive that such a man, a man who had such deep beliefs that we can call anthropological, or better genetic, using a modern term, could change his heart to follow the wind of the moment that, more or less, is the art of the politicians? If he had been a 'good politician', he could have become the 'Chaplain of the Empire', the mouthpiece of Caesar 'the tyrant', i.e. the mouthpiece of tyranny. This is, on the part of the historians here above mentioned, not an error, that they should well have avoided, but a heresy.

Morkel answers Cicero's critics by writing that the great orator was

> "kein Mundstück des Servilismus (Mommsen) ... er hat Caesar mitunter arg geschmeichelt, aber viel häufiger hat er direkt oder indirekt attackiert ... er hat den Mut, für seine Überzeugungen einzustehen ... war er kein Opportunist ... das Ziel stand für Cicero immer fest: die Rettung der Republik",

he was

> "no mouthpiece of servility (Mommsen) ... he flattered, badly, Caesar from time to time but much more frequently he attacked him directly or indirectly ... he had the courage to take responsibility for his convictions ... he was no opportunist ... his aim was always the same: the salvation of the Republic".[52]

And, last.

> "Den Kampf um die römische Republik hat er verloren, im Kampf für die Idee der Republik sind seine Schriften bis heute eine unversiegbare Quelle geblieben", "he lost

---

[51] See above, 'Mommsen and Cicero', text coresp. with n. 31.
[52] Morkel, op. cit., p. 285.

the battle for the Roman Republic but, for the struggle for the idea of the Republic, his writings have remained until today an inexhaustible source".[53]

Wolfgang Schuller's aim, in his *Cicero, oder der letzte Kampf um die Republik*,[54] is that of "Ciceros Leben so anschaulich darzustellen wie möglich", that of portraying Cicero's life as clearly as possible, and this aim is fully attained by the author. The narrative is pleasant, and portraits and maps intersperse the chapters which, in this way, adds to its readability. Obviously, Mommsen is substantially ignored. The subtitle "der letzte Kampf um die Republik", the last battle for the republic, is also indicative of the author's point of view, that is that of the most recent, mature, German historiography.

Francisco Pina Polo's *Marco Tulio Cicerón*[55] is reviewed in this section, along with the other German books on Cicero, because, although the work of a Spanish scholar, as said above, it has been substantially *adopted* by the German world of learning. It has in fact been published in a German translation c/o Klett-Cotta Verlag in 2010, and soon after, in 2011, has undergone a second edition. This proves how clearly it responds to the wish of the cultured public, in Germany, to know more about Cicero. The book is essentially a political biography,[56] and some of its author's points are extremely modern, as for concepts and wording, proper to the present political debate. For example, when he writes that Ciceronian political thought "in some respects is an intimation of the basic lines of modern liberalism", or when he speaks of "unnecessary state's interventionism instead of the more desirable private initiative".[57]

The substantial idea that leads Pina Polo is that Cicero was a Republican Conservative, "un republicano conservador", and beyond this judgment it is apparently difficult to move. Nevertheless biog-

---

[53] Ibid., p. 289.

[54] Beck Verlag (München, 2013).

[55] Editorial Ariel (Barcelona 2005). See above, 'Mommsen and Cicero', n. 29.

[56] " ... se centra sobre todo en su vertiente política, que constituye el hilo argumental del libro", ibid., pp. 10-11.

[57] "El pensamiento ciceroniano ... prefiguraba en algunos aspectos las líneas básicas del liberalismo moderno ... un intervencionismo innecesario del Estado frente a la más deseable iniciativa privada", ibid., p. 55.

raphers and historians of ideas often conflict in their judgments, and also in connection with this book some reflections appear necessary.

The Arpinas was undoubtedly a man of his own times, with his own beliefs and his own ideas concerning the organization of society. The Aristotelian concept of social hierarchies had not been questioned and he, the *homo novus*, was proud of his achievements and, at the same time, conscious of his own merits, while traditional nobility remained somehow prisoner of her own past. Nevertheless we cannot think that 'historical change' was unknown to Roman society. The emergence of new ranks was in fact an everyday matter, because no institution can prevent it, as no institution in history has ever been able to prevent it for long. And Roman nobility was obviously impoverished and less influential than in the past, and Cicero was himself a *homo novus*, a new man in his own society.

Cicero played his own role as a man of distinction, and his ideas, as was inevitable, were somehow influenced by the station he had in life. Nevertheless his concept of the mixed constitution meant also that the popular element no less than nobility and intermediate ranks had its own role to play in society.

For the rest, it seems rather obvious that, from his point of view, the *optimates* should have ruled the Republic, while his *concordia omnium bonorum* makes the modern reader think of the *Patriot King* of Lord Bolingbroke, and of the ethics that it embodies.[58]

Pina Polo describes Cicero as a conservative politician, unable to understand the necessity of social reforms.[59] This can be true from a biographical point of view, but to defend the Republic meant also

---

[58] A. Ward, in his *The Tory View*, op. cit., writes that Bolingbroke "finds in Cicero's Rome an exact parallel to Georgian England". In particular, Bolingbroke wrote: "We must not imagine that the freedom of the Romans was lost because one party fought for the maintenance of liberty; another for the establishment of tyranny; and that the latter prevailed. No. The spirit of liberty was dead, and the spirit of faction had taken its place on both sides", Ward, ibid., p. 429.

[59] The concordia ordinum and the consensus omnium bonorum "proponían una gran coalición de proprietarios para defender, mediante un férreo control del poder político, sus intereses privados frente a las demandas sociales ... Pero, para una persona con la mentalidad de Cicerón, era casi imposible aceptar la necesidad ... de introducir reformas que permetieran la supervivencia de la República", ibid., pp. 252-3.

to defend liberty and all the changes that, in society, take place in a regime of liberty.

Certainly, a parallel with Socrates is hard to draw. Socrates is an idealized figure, whom we know thanks to Plato, and he is an example to admire especially in the *Apologia*, on his refusing to escape from prison to avoid setting a bad example to youth. Cicero is a terribly human being, a man of his own times, whom we know better than any other man of the Roman Republic thanks to his letters. Nevertheless he is grandiose in his defence of the Republic, to which he sacrificed his life.

Mommsen could well write that in his dialogues he was inferior to Plato. Nevertheless his versatility as a writer is unequalled, and in some passages of his philosophical works highly suggestive.

On some recent books

Finally, in chronological order, John Dugan, *Making a New Man. Ciceronian Self-Fashioning in the Rhetorical Works,*[60] Henriette van der Blom, *Cicero's Role Models. The Political Strategy of a newcomer,*[61] along with Stephanie Kurczyk,[62] develop an analogous concept, which seems now to whet the appetites of young researchers.

Dugan's book evolved from a doctoral thesis, and aims at examining the ways in which both Cicero and other Roman writers describe the 'self', especially that of a new man. It emphasizes the notion that this description is a product of deliberate strategies of fashioning, and intends to analyze these strategies within Cicero's rhetorical thought.

The author refers to other scholars, who have discovered the subject of self-fashioning, especially to Stephen Greenblatt and to his *Renaissance Self-Fashioning. From More to Shakespeare.*[63] Greenblatt, he writes, has provided his investigation "with the basic methodological framework for its study of how authors construct their identities within literary texts". But also Emanuele Narducci,

---

[60] Oxford U. P., (Oxford, 2005)
[61] The Oxford Classical Monographs (Oxford, 2010).
[62] See n. 37 and text corresp. with notes 37-9.
[63] Univ. of Chicago Press (Chicago, 1980).

in his *Cicerone e l'Eloquenza Romana*,[64] as Dugan adds, analyses Cicero's *rhetorica* as constituting a comprehensive educative project. Therefore, "Cicero's rhetorical writings work together to form a coherent cultural programme".[65]

Van der Blom's aim is that of examining "Cicero's rhetorical and political strategy in late Roman Republican politics with regard to his self-advertisement as a follower of chosen models of behaviour from the past, his role models as personal *exempla* ... Cicero presented himself as emulating specific historical figures with the purpose of building up and strengthening his public persona and thereby supporting his claim to political offices and influence".[66]

The book is well structured, well written, the result of a long and detailed study. The problem is whether Cicero had a conscious strategy, in presenting himself as emulating etc. etc., or not. Authors who tend towards hagiography would be tempted to answer no, but Van der Blom's demonstration is quite convincing, quite helpful in knowing better Cicero the man and the orator. Therefore, from this point of view it is a welcome addition to historical literature.

James May (ed.) with his *Brill's Companion to Cicero. Oratory and rhetoric*,[67] as we see from the title, aims to settle the problems of oratory and rhetoric in Cicero.

Judgments on rhetoric raise problems as for Cicero's intention and his behaviour in politics. For example, concerning the *Pro Marcello*, Harold Gotoff, one of the contributors, observes that "early critics read it as a *gratulatio*, focusing on what they took to be Cicero's servile adulation of Caesar, but now scholars focus on the speaker as a political advisor".[68]

While praising Caesar, Cicero has in fact in mind the return to republicanism. He glorifies Caesar but, at the same time,

> "he asserts his independence and frankness ... he advises him of the need to take swift action to repair the republican system ... thus by publicly challenging Cae-

---

[64]Laterza (Bari, 1997).
[65]Dugan, op. cit., p. 16.
[66]Van der Blom, Preface, p. IX.
[67]Brill (Leiden, 2002).
[68]H. Gotoff, 'Cicero's Caesarian Orations', ibid. p. 226.

sar, Cicero maintains his persona as an independent adviser".[69]

With regard to the *Philippics*, Jon Hall writes that "Cicero's powers in this final stage of his career ... show no sign of stagnation or decay ... Perhaps the quality that stands out most of all in these speeches is their vigor ... Cicero is at least his own man, and his oratory is all the better for it. His formidable oratorical powers are finally directed towards a clearly defined cause in which he fully believes".[70]

Other works on Cicero include Catherine Steel, *Cicero. Rhetoric and Empire*,[71] in which the author emphasizes the point that Cicero was the only eminent Roman who chose not to spend time outside Italy. Apart from some periods away from the city, i.e. his early trip to Rhodes and the time in Sicily, "his absences were forced upon him". So "exile, the province of Cilicia and the pursuit of Pompeians during the civil war ... it is in these periods ... that Cicero the orator is replaced by Cicero the letter-writer".

This notwithstanding, a large number of Cicero's speeches deal directly with issues arising from Rome's possession of an empire, and the aim of Steel's book is "to examine Cicero's analyses of imperial problems".[72]

The concept of *imperialism* of the Romans, suggested by the author, demands comment.

Certainly, when seeing Roman ruins all around Europe, from Scotland to Spain, from western Germany to eastern Europe, from northern Africa to the Middle East, one can scarcely not admit that the Romans were imperialists, at least in the sense, quite modern, that is usually given to this term. The history of the Roman Republic, afterwards of the Roman Empire, is in reality a sequence of wars, with the exception, in part, of the conquest of Britain, fought

---

[69]Ibid., pp. 231-2.
[70]Jon Hall, 'The Philippics', ibid., p. 302.
[71]Oxford UP (Oxford, 2001). By the same author see *Reading Cicero. Genre and Performance in Late Republican Rome*, Bloomsbury (London-New York, 2005) and, with Henriette van der Blom, (eds), *Community and Communication: Oratory and Politics in Republican Rome*, OUP (Oxford, 2013).
[72]Ibid., pp. 2-3.

for defence, or in the pursuit of security. This is Machiavelli's concept, according to whom, in the vain search for security, states fight until they fall into ruin.

So the conquest of Gaul, that we could only in part ascribe to an imperialist concept. In reality Gaul was a nightmare for the Romans, since the first years of the Republic (see the legend of the geese of the Capitol that alarmed the sleeping defenders against the Gauls who, at night, were assailing the rock). In 102 BC at Aquae Sextiae, near Aix-en-Provence, in Southern France, Gaius Marius defeated the Cimbrics, who were about to invade Italy. The next year he had to face the Teutons, who had crossed the Alps, and defeated them at the Campi Raudii, near Vercelli, in Piedmont. When Caesar finally conquered Gaul, there was a general sense of relief in Rome.

Therefore, the history of Roman conquests in Europe seems to be ascribed to different factors. Sometimes for self-defence, sometimes for the spirit of conquest, that leads men in their actions, -this is again a Machiavellian concept- they found themselves the masters of Europe, to which, anyway, they brought civilization.

The history of the civil wars, for example, can also be explained using only Machiavelli's anthropology, that depicts man as a violent and selfish being, the prey of instincts, substantially similar to the beasts, whose nature he shares.

The end of the republic, when power was in the hands of men of war, while Cicero was only an unarmed prophet, recalls to mind the long debate, in seventeenth and eighteenth-century Britain, on the standing army, that the liberals considered as an instrument of despotism.[73]

And power, in Rome, in Cicero's times, was no longer in the hands of Senate, but in the hands of commanders of armies who, therefore, sealed the end of the Republic.

Shane Butler, in his *The Hand of Cicero*,[74] sets out to demonstrate the direct relationship between oratorical practice and writ-

---

[73]See, for example, John Millar of Glasgow, *The Historical View of the English Government*, in four vols., (second ed. London, 1803), who largely debates this issue.

[74]Routledge (London, 2002), pp. IX-165.

ten text in Rome, illustrating, in particular, "the synchronic relationships between written and oral practice".[75]

Matthew Fox, *Cicero's Philosophy of History*,[76] is a 'provocative' book, as the author defines it, since philosophy of history at the time of Cicero did not exist. In part under the influence of the Cambridge historians, that is detectable in the tissue of the narrative, Fox explores what he calls Cicero's engagement with history, that he considers as "profoundly philosophical", and maintains that this "will attract a variety of readers to explore in Cicero's writings a constellation of ideas that has not so far been tackled directly in existing scholarship".[77]

Fox complains about the neglect of the scepticism of Cicero's Academic training, and his clear insistence (as in the opening of the *De Natura Deorum*) that "philosophy consists in making a range of ideas available rather than insisting on the supremacy of one particular doctrine".[78]

On this Ciceronian concept the present author fully agrees, convinced, as he is, that philosophy is *problematic*, in the sense that its aim is that of raising and solving problems, not that of affirming a particular doctrine. In any case, unless it has, as repeatedly said above, as its background the idea of the 'whole', or the Hegelian *Gesamtheit*, it cannot be considered as true philosophy, but just as an ideology, to serve the interests of a particular social group.

Fox's analysis in the chapters on the Enlightenment, when he maintains that "Cicero's reputation was, to say the least, controversial", is particularly engaging.[79]

This judgment nevertheless invites reflection, if considering what the young Montesquieu, Adam Smith, Adam Ferguson, David Hume himself, wrote about Cicero, or how they repeated from his works not only the concepts but often even the very wording. And to the Scots must be added the admiration of Friedrich der Große and Voltaire, among others.

With the eighteenth century we certainly observe "the shift of philosophy into something as a distinct profession", which became

---

[75]Ibid., p. 3.
[76]OUP (Oxford, 2007).
[77]Ibid, Introd., p. 1.
[78]Ibid., p. 8.
[79]Ibid., p. 73.

firmly established particularly with Hegel. Nevertheless, the present author maintains, with regards to eighteenth-century cosmology, it needs to be demonstrated that it had made substantial advances on the Stoic and Ciceronian one, whose basis was only in part superseded by the new discoveries.

As for human societies, the historian of science will emphasize what is new in scientific discoveries, the historian *tout court* will inevitably insist on the survival of ancient ideas. Nevertheless it seems clear that, notwithstanding the new advances in science, much of Ciceronian cosmology was still alive in that century.

No doubt, there was no idea of force in the ancient world. To be sure, for the Newton of the *Optics* the forces have multiplied, to include capillarity, cohesion, chemical bonds & repulsions, electricity and magnetism,[80] and all this was unknown to the ancients. Furthermore, the concept of philosophy as a distinct 'profession' became more firmly established, in part with Hume, definitively with Hegel, but not before the end of the century, and Hegel is a man of the early 19th century rather than an eighteenth-century man. He is the man of Romanticism, not of the Enlightenment. And how much of the new physics can we find in the works of Hume, Smith, Ferguson, and of the other writers of that century? Not much, indeed.

The eighteenth century is a long century. Concerning Cicero, it begins with John Toland's *Cicero Illustratus*, of 1712.[81]

Toland's main claim to fame, as Fox writes, was as the coiner of the term 'deist', and he was the promulgator of a philosophical kind of religion, divested of any traces of superstition. His project was an edition of Cicero's complete works, including indices, introductory essay and explanatory notes, a project for which we had to wait until the edition by Johann Caspar von Orelli, of 1826-1838.

But Cicero was a multifaceted writer, and inevitably each age has found and emphasized in him particular aspects of his doctrine. If nowadays we can see in him, and this will probably be judged as a heresy, not only a precursor, but a complete theorist of modern

---

[80]These words are by the distinguished Canadian colleague Roger Emerson, in a private communication.

[81]Its full title is *Cicero Illustratus, Dissertatio Philologico-Critica: sive consilio de toto edendo Cicerone, alia plane methodo quam unquam factum*, John Humphreis (London, 1712).

liberal societies, in the eighteenth century he could well appear as a deist, let us say as the theorist of a religion without priesthood and without superstition.

Matthew Fox's book, so engaging in its reasoning, is a welcome addition to historical literature, since it explores aspects of Cicero's doctrine until now not sufficiently emphasized. Nevertheless we emphasize the point that, notwithstanding the obvious advances in scientific research, the bulk of Cicero's doctrine concerning cosmology was not superseded until the end of the eighteenth century, and the concept of force of gravity will never be superseded, at least for the next few centuries, if not for millennia to come.

Andrew Lintott, *Cicero as Evidence*,[82] can be defined as a critical reading of Cicero's works, mainly of his oratorical works, and is the result of forty years of teaching Cicero, as the author says in the preface.

He begins by observing that "one of the first things to learn" on the part of students of late-Republican history, is that they "cannot treat Ciceronian texts as authentic records of history ... Cicero is not a detached and impartial narrator of either the world in which he himself moved or the past history of Rome ... In the courts of the Roman Republic an orator's duty was to his client, not to the court".[83]

Moving from these premises, Lintott detects much evidence in Cicero, explains much that is missing from his speeches, and tries to explain the reason why it is missing.

For example, Cicero was frequently one of a team of orators, often speaking last, when emotional appeal was more important than argument from evidence. Therefore, we do not know all the arguments on his side, "let alone what was said by the opposition".[84] And Quintilian says that busy orators usually improvised the greater part of their speeches, only writing the beginnings and their most essential parts.

Therefore: we possess Cicero's speeches as a coherent oratorical genre, but what did actually happen in the courts?

---

[82](Oxford UP, 2008).
[83]Ibid., p. 1.
[84]Ibid., p. 19.

Moving from these premises, Lintott proceeds to an examination of Cicero's speeches. In this way he marks a step forward in the understanding of the *Orationes*, beyond purely literary appearance. Taking at face value anything the orator published, could in fact be misleading, while one must read with a critical eye, to detect what he did not say.

Although this book is "more of an intellectual history of the man than a history of his actions",[85] yet it cannot be ignored by the historians of Cicero's life and of the history of the late Roman Republic.

Kathryn Tempest, *Cicero. Politics and Persuasion in Ancient Rome*,[86] more than dealing with 'politics and persuasion', is in the reality a biography, quite sympathetic with Cicero as a man and politician. The author admits that her book has probably "gone too far in defending his motives and actions", but this is "because this position has been too frequently overlooked".[87]

The book makes, in any case, good reading, the narrative is well ordered, the portrait it gives of Cicero is honest, balanced, fairly dealt with. From it Cicero comes out as a human being, with all the weaknesses and attributes of a human being.

Someone could observe that the book does not always reach a critical stage, nevertheless the judgment of its author comes out naturally from the narrative, and does not necessarily need to be widely discussed, or justified. It is in any case a welcome addition to the bibliography on the subject, a book that someone had to write.

The most recent *The Cambridge Companion to Cicero*,[88] edited by Catherine Steel, settles the problem of Ciceronianism and of its meaning to the present. It is divided into three sections, the first on Cicero as the 'Greco-Roman Intellectual', the second on 'the Roman Politician', the last on 'receptions of Cicero'. It is an introduction to the subject, a necessary reading for those who want to go further in Ciceronian studies.

---

[85] Ibid., preface, p. V.
[86] Continuum Publishers (London and New York, 2011, repr. 2014).
[87] Ibid., p. 210.
[88] C. Steel, ed., n. 1, n. 71.

Last, *The Classical Journal*, No. 1, Oct.-Nov. 2014, Special Issue on Cicero, pp. 1-128, needs to be mentioned, to conclude this section.[89]

Again last, but not least, in the city of Arpino, along with the *Certamen Ciceronianum Arpinas*, which takes place every year in the month of May, also a symposium takes place, to which several scholars, mainly Italian, contribute. Emanuele Narducci ran them for years, until his untimely death, in 2007. The symposia are now run by other scholars, but they deal almost exclusively with classical philology, only marginally contributing to the history of ideas.[90]

Political thought

There is not much, indeed, in the bibliography on Cicero's political thought. Apart from Friedrich Cauer and his *Ciceros Politisches Denken*,[91] a book now superseded, although it had its own *raison d'être* when it was published, and apart from Hermann Strasburger, *Concordia Ordinum. Eine Untersuchung zur Politik Ciceros*,[92] only Neal Wood has extensively dealt with the subject in his *Cicero's Social and Political Thought*.[93]

Hermann Strasburger complains that scholars have continued to write "die Vorstellungen von der Existenz ... politischer Parteien in modernem Sinne (Optimaten, Populare) die es in Rom gar nicht gab",[94] "the representation of the existence of political parties in the modern sense ... that in Rome did not exist at all". He then maintains that the concept of *concordia ordinum*, or harmony of the ranks, i.e. the Greek ὁμόνοια, was "as old as the Republic", and

---

[89] Other books include Sarah Culpepper Stroup, *Catullus, Cicero and a Society of Patrons*, (Cambridge UP, 2013), and, from Bloomsbury Publishers: David Taylor, *Cicero and Rome*, 2013; Thomas Wiedemann, *Cicero and the End of the Roman Republic*, first published in the Bristol Classical Press, (1998), now in the Bloomsbury series.

[90] See, in any case, the series *Ciceroniana*, a collection of *Acta* in several volumes, edited by Scevola Mariotti, c/o the 'Centro di Studi Ciceroniani', Piazza dei Cavalieri di Malta 2, 00153 Roma.

[91] Fr. Cauer, *Cicero's Politisches Denken* (Berlin, 1903).

[92] First published in Borna, 1931, Fotomechanischer Nachdruck (Amsterdam, 1956).

[93] University of California Press, 1988.

[94] Ibid., Vorwort, p. IV. He apparently refers here to Mommsen.

Cicero gave it "theoretische Fundamentierung ... in seinen staats-philosophischen Schriften".[95]

Nevertheless, it seems that this concept never became philosophical in Cicero, and remained just a political one, a mere exhortation to concord, of which we have so many examples in history, in critical moments of human societies (see, for example, Lord Bolingbroke's *The Patriot King*).[96]

The episode of the Gracchi brothers is meaningful in itself, and seems rather to justify the Marxist historians, when they see, in societies, the opposition of capitalists and workers. In their view, the substance of the matter is that there were social classes in Rome, although an ideology had not yet been theorized. And an ideology, according to them, is a mere *superstructure*, which is elaborated later, as a mere reflex of the matter, while the substance of history, or its structure, is in the division of societies in classes.

The question that arises in relation to Strasburger's work is, why human societies have always structured themselves in the same way, like a pyramid, with very few people at the top and many at the base. So are societies nowadays, and so was the Roman society of Cicero's times. Again, the question arises: should we just try to *explain how* these societies historically were? But this could mean to substantially *justify* them. Or should we condemn the societies of the past as a capitalistic aberration, to be superseded with the help of class struggles?

An answer depends on the general views of a researcher, on the premises of his reasoning. Marxist scholars will inevitably see societies as structured in classes, Liberals will see in them only ranks, mobile and changeable. The opinion of the present author, who sides with Adam Smith in political economy and in the concept of natural society, is that, from a philosophical, and also from a political point of view, both the theories contribute to the *knowledge* of human societies and to their progress. Mankind, therefore, needs both.

Strasburger was a distinguished classical philologist, and his research is well documented and well developed. Although, as was inevitable, he did not attain the level of political philosophy, which

---

[95] Ibid., p. 12.
[96] Cited above, n. 58.

was beyond his own interests, yet with the *concordia ordinum* he explained a fundamental concept in what we can call Cicero's system of political ideas.

Marxism is the lantern that leads Neal Wood in his *Cicero's Social and Political Thought*, where the author intends to demonstrate no more and no less than Cicero was a defender of private property.

He surprises the reader by beginning with a real paean to Cicero, writing that

> "the peak of (his) authority and prestige came during the Enlightenment ... it was a Ciceronian century ... Cicero was a leading culture hero of the age: revered as a great philosopher and superb stylist, hailed as a distinguished popularizer, and praised as a humanistic sceptic who scourged superstition; a courageous statesman and dedicated patriot, the ardent defender of liberty against tyranny".

Voltaire, Montesquieu, Diderot, the French revolutionaries, all praised his scepticism, republicanism and libertarianism. In Great Britain Gibbon, Johnson, Pitt, Fox, Sheridan, were his admirers, and Hume and Smith were particular ones, but no one was a more devoted Ciceronian than Edmund Burke. And so in America, where the founding fathers were inspired by him.

Wood then complains that

> "both conservative and radical thinkers seem to have taken from him what they wished to underpin their differing positions, ignoring the more uncongenial aspects of his thought ... they selectively exploited ... the principles of natural law and justice and of universal moral equality; a patriotic and dedicated republicanism; a vigorous advocacy of liberty, impassioned rejection of tyranny, and persuasive justification of tyrannicide; a firm belief in constitutionalism, the rule of law, and the mixed constitution ... a moderate and enlightened religious and epistemological scepticism".[97]

---

[97] Ibid., p. 4.

Nevertheless in the nineteenth century, with the rapid economic and demographic changes in Western Europe, the growing industrialism and urbanization, we witness the "downfall and discredit" of Cicero, who then appears as *"the sworn enemy of popular rule, the implacable foe of social amelioration and economic reform, a leader of the Roman landed oligarchy who decried any drift toward arithmetical equality or social parity*[98] ... The young Marx wrote in the *Notebooks on Epicurean Philosophy* (1839) that he "knew as little about philosophy as about the President of the United States of America". "The landmark of this tendency was Theodor Mommsen's *Roman History* (1854-56) a detailed and sweeping analysis ... dealing a final blow to the prestige of Cicero ... Cicero's reputation has never recovered from the stress and shifts in fashion of the nineteenth century, despite such attempts to rehabilitate him as Zielinski's classic reply in 1912 to both Mommsen and Wilhelm Drumann. Today Cicero is seldom taken very seriously except by classicists". In particular, Sir Frederick Pollock writes: "Nobody that I know has yet succeeded in discovering a new idea in the whole of Cicero's philosophical and semi-philosophical writings".[99]

For the present author, neither Karl Marx nor Sir Frederick Pollock were demigods, and Mommsen certainly was not the 'dedicated liberal' that Neal Wood, as he calls him, believes that he was.

Therefore, why is Wood praising so highly Cicero? Why is he complaining that Marx, Pollock, Fr. Cauer, G. H. Sabine and others, with their works "are ample testimony to a depressing aspect of intellectual life: the sterile repetition from generation to generation of a stereotyped interpretation of a specific thinker without deviation or spirit of critical inquiry"?[100]

The reason is that Cicero was

> "the first major social and political thinker of antiquity to offer a formal definition of the state. He was also the first to stress private property ... and the importance of the state for its protection ... For Cicero the state exists primarily to safeguard private property and the accumulation of property ... He was the first major

---

[98]Emphasis mine.
[99]Wood, ibid., pp. 6-8.
[100]Ibid., p. 9.

social and political thinker to distinguish clearly state from government ... to separate conceptually state from society".

In other words, in Wood's conception, Cicero is the theorist of the state, considered, in Marxist terms, as an instrument of oppression of the privileged classes against the inferior ones. Therefore, against Pollock's judgment, he is sure that, for this particular aspect of his theory, if Cicero is decidedly "not one of the greatest social and political thinkers of our culture ... he is certainly entitled to a place with major political thinkers of the second rank: Machiavelli, Hume, Bodin, Montesquieu, Burke and J. S. Mill".[101]

Marxist premises, as said above, lead Wood in his analysis of Cicero's work and of Roman society, and one needs to admit that, if seen through the lens of that theory, they reveal aspects that should be emphasized, while, instead, they have mostly been neglected by historians.

For example, he writes that the whole nobility were the 3% of the Italian population controlling the destiny of 50 million souls in the Empire.[102] When consul, in 63, Cicero "successfully blocked any movement for reform, safeguarding landed and business interests, preventing the reduction and cancellation of debts";[103] his was a "tacit aristocratic assumption of proportionate equality: more is owed to the superior ... less to the inferior ... (this) favours the privileged few to the detriment of the unprivileged majority ... a conservative who wishes to justify the maintenance of a hierarchical and authoritarian system of political and social inequality dominated by a wealthy landed class";[104] "society which he cherishes ... hierarchy rank and order ... very little sensitivity and compassion for the humble people, contempt for the lower orders, vulgus or multitudo".[105]

Wood teaches us something new about Cicero, less about the concept of society. He admits that in Rome "violence was not ...

---

[101] Ibid., pp. 11-12.
[102] Ibid., p.15.
[103] Ibid., p. 49.
[104] Ibid., p. 76.
[105] Ibid., p. 94.

of a revolutionary nature".[106] Catiline and Clodius were motivated more by honour than by social reasons, and bankrupt aristocrats were the worst demagogues. Furthermore, Cicero was "a master of the practical art of politics who would have had little to learn from Machiavelli about the acquisition, conservation and increase of power",[107] because he describes *populares* as "mentally deranged, morally lost, and insolently reckless criminals and incendiarists".[108] In sum, in Wood's view, in his contempt for the lower orders Cicero was Machiavelli's precursor.

One could answer recalling the attention of the reader to the violence, that was the main character of Roman society during the last period of the Republic. The point is, furthermore, that Aristotle, St. Thomas, Locke, never challenged the type of society as it had constituted itself during the centuries, or the millennia, and they took inequality for granted.[109] The problem, we repeat, is that of answering the question, why societies of the past structured themselves in that particular way, and not in a different one. After Cicero's death and the fall of the Republic, things did not improve from that point of view but, possibly, worsened, with the addition of tyranny, that sealed the fate of hundreds of millions of people for two millennia, until the struggles for liberty in seventeenth-century England took place. Even these struggles did not substantially change the structure of society, where men continued to be moved by the same motivations or, if one prefers, impulses, having recourse to Machiavellian vocabulary.

It should now be clear that there are ranks in society, not social classes, and that classes can be imposed only with the force of

---

[106] Ibid, p. 35.

[107] Ibid. p. 177.

[108] Ibid. p. 195.

[109] For example, one should consider how long it took, in the late 'liberal' eighteenth-century Britain, to assert the concept of progressive taxation. Concerning this long lasting debate see John Millar, *Letters of Sidney on Inequality of Property*, appearing in *The Scots Chronicle* in 1796. This concept is now generally accepted in Europe, less so in the United States. And the United States is 'the oldest democracy in the world', as somebody observed, some time ago.

Millar's letters are now reprinted in V. Merolle (Ed.), *Letters of Crito and Letters of Sidney*, Giuffré (Milano, 1984).

coercion.[110] Even if not having behind him two or three millennia of history, about which to reflect and learn, as a theorist Cicero was extremely modern in his fight against tyranny and in his concept of the mixed constitution. And the mixed constitution was no more and no less than the modern concept of liberty. Concerning it, Cicero writes in the *De Republica*:

"For just as in music ... a certain harmony of the different tones must be preserved ... and as this perfect agreement and harmony is produced by the proportionate blending of unlike tones, so also in a state made harmonious by agreement among dissimilar elements, brought about by *a fair and reasonable blending together of the upper, middle, and lower classes*, just as if they were musical tones. What musicians call harmony in songs is concord in a state".[111]

Certainly, Cicero was inevitably a man of his own times, but everybody recognizes that, if considering the nature of his own society, he was one of the best people one could hope to come across.

Therefore, one cannot demonstrate that he was an inspirer of Machiavelli, as Wood maintains.

As for Machiavelli, as said above, to understand him one must begin from his anthropology, that considers men as wicked, as the prey of material needs, moved to evil by their own constitution, that is substantially similar to that of the beasts. Man is therefore exactly the opposite of the man 'made as the image and likeness of God' of traditional religion. One may well wish that mankind were morally better, but some thousands of years of history, if considering even the wars that are waged nowadays, while reason should prevail, prove that Machiavelli was not wrong at all.

After all, in ancient Rome the current motto was *panem et circenses*, i.e., to the plebeians must be given corn and games. And the Florentine thinker, fifteen centuries later, had seen bitterly his fellow-countrymen prey of the superstition, when believing in

---

[110]See in particular the classical *The Origin of the Distinction of Ranks*, by John Millar of Glasgow, first published in 1771.

[111] "Ut in cantu ipso ac vocibus concentus est quidam tenendus ex distinctis sonis ... isque concentus ex dissimillimarum vocum moderatione concors tamen efficitur et congruens, sic ex summis et infimis et mediis interiectis ordinibus ut sonis moderata ratione civitas consensu dissimillimorum comcinit; et quae harmonia a musicis dicitur in cantu, ea est in civitate concordia", *De Rep.*, II, xlii, 69. Italics mine.

the prophecies of Fra' Girolamo Savonarola.[112] Hence his reflections about the *vulgus*, or the vulgar, quite distinct from *color che sanno*, or those who know.

Furthermore, slavery was justified by Aristotle and by all thinkers of antiquity, nor did the Romans challenge it. Why should Cicero have challenged it? He was an orator, a politician, a philosopher, not a prophet. And yet he fought to the last for republican ideals, or for ideals of liberty, which were the forerunners of the liberation of mankind.

It is therefore a paradox, on the part of Neal Wood, who was an iconoclast, a desecrator, although not without merits, to maintain that Cicero was Machiavelli's precursor, a Machiavellian *ante litteram*. But Machiavelli was a materialistic writer. His political philosophy certainly does not support Wood's judgment. To understand Machiavelli, one must think of his 'overturning of the traditional way of thinking'.

What does this mean? Simply: that before him all the philosophers in their speculations had moved from the existence of God, of the God of traditional religion, as the Creator of the world. Machiavelli overturns this way of thinking. He moves from the concept of the man in actual fact living and acting, and ignores religion altogether. And religion is for him only an *instrumentum regni*, an instrument to reign, mostly in the hands of tyrants. This is the fundamental concept to understand Machiavelli, a concept quite alien to Cicero, the Pagan Christian, as Petrarch calls him.

---

[112] A picture quite similar is in Cicero, about the situation in Italy, in the year 49, during the First Triumvirate: "Pompey, a hopeless failure as a statesman, and now I find him an equally bad general ... He says the optimates are tearing me to shreds. What sort of optimates, in heaven's name? Look at the way they are now going out to meet Caesar and positively currying his favour. As for the towns, they make a god of him, ... the truth is that any evil this Pisistratus had not done is earning him as much popularity as if he were to have stopped someone else doing it. In him they hope to find a gracious power ... you can imagine the town deputations and official compliments. You will say they are frightened. I dare say they are, but I'll be bound they are more frightened of Pompey than of Caesar. They are delighted with his artful clemency ... so ask you, what sort of optimates are these that thrust me out while they themselves stay at home?", Ad Atticum, letter 166, Formiae, 4 March 49.

On Mommsen

We have cited above Mommsen's biographers, Ludo Moritz Hartmann, *Theodor Mommsen. Eine biographische Skizze*,[113] Alfred Heuß, *Theodor Mommsen und das 19. Jahrhundert*,[114] Lothar Wickert and his four-volume *Theodor Mommsen. Eine Biographie*[115] and, last, Stefan Rebenich, *Theodor Mommsen. Eine Biographie*.[116]

Although there is more, in Germany, on this subject,[117] Rebenich's biography is the most comprehensive study of Mommsen's life and works, and has benefited from the 'fall of the wall', which rendered accessible documents and books that were located in libraries and archives in the DDR, the former Deutsche Demokratische Republik.

Rebenich's book is fascinating reading. When, for example, it describes Mommsen's first arrival in Rome, travelling from Paris via Genua and Florence, it recalls to mind of the reader Gibbon's words when, sitting on the Capitol, the great English historian first conceived the idea for his *Decline and Fall of the Roman Empire*.[118]

The book well reconstructs the cultural environment of Germany in the first part of the nineteenth century, that cultural environment

---

[113] Perthes (Gotha, 1908).

[114] Franz Steiner Verlag (Stuttgart, 1996), Nachdruck der Ausgabe Hirt, Kiel, 1956.

[115] In 4 vols, Klostermann (Frankfurt am Main, 1959-1980). Wickert is severely criticized by Rebenich, who writes that, apart from relevant omissions, "dem konservative Wickert war der liberale Mommsen völlig fremd ... Wickert ... hat nicht nur die Geschichte des 19. Jahrhunderts nicht verstanden, sondern auch den Gegenstand seiner jahrzehntenlangen Forschungen: Theodor Mommsen" "to the conservative Wickert the liberal Mommsen was quite alien ... Wickert did not understand not only the history of the nineteenth century, but also the object of his decades-lasting research: Theodor Mommsen", Rebenich, p. 229.

[116] (München, 2002, in der Beck'schen Reihe, 2007).

[117] See in particular Fritz Sturm, *Theodor Mommsen. Rückblick auf Leben und Werk* (Karlsruhe, 2006).

[118] "It was at Rome, on the fifteenth October 1764, in the close of the evening, as I sat musing in the Church of the Zoccolanti or Franciscan friars, while the barefooted friars were singing Vespers in the Temple of Jupiter on the ruins of the Capitol, that the idea of writing the decline and fall of the city first started in my mind", Gibbon, *Memoirs of my Life*, edited by Betty Radice (London, 1991), p. 148.

from which the young Mommsen so well benefited, along with the political struggles in which he took part.

Rebenich, as any biographer, risks falling into hagiography in the defence of his hero, but it is a risk he is able to avoid. For example, he admits that the *Römische Geschichte* was substantially a diversion from Mommsen's great philological work, when insisting that his masterpiece is the *Römisches Staatsrecht*.

He gives the right emphasis to the criticisms that the *Römische Geschichte* met, particularly by Nietzsche, Droysen, von Treitschke. None of them were answered by Mommsen. But, when insisting on Mommsen's liberalism, the unresolved contradiction strikes the reader, who wonders, how can it be that Mommsen's "Vorstellung eines parlamentarischen Systems wurzelte in der politischen Philosophie Benjamin Constants", that "Mommsen's idea of a parliamentary system had its roots in the political philosophy of Benjamin Constant"?[119]

Constant de Rebecque, or Benjamin Constant, as he was called, was the head of the parliamentary opposition after the Restoration, but during Napoleon's despotism he was repeatedly in exile, as all true liberals were.[120]

And, why praise Napoleon III, the new Caesar, while Tocqueville, the other great French liberal writer of the nineteenth century, was in prison, although only for a short period, for opposing the new despot?

Our conclusion, as demonstrated above, is that Mommsen was a liberal -not uncompromising, indeed- in the main political struggles of his years in Germany, but philosophically he was not so. We can call him a liberal in an immanent sense, not in a transcendent one.

Therefore, politically he was mostly a liberal in his opposing Bismarck, von Treitschke, the *ostelbischen Junker*, and in other struggles, although with too many oscillations, but philosophically he was not so, as shown in his praising the tyrants, Caesar and Napoleon III.

In sum, he was an actor on the stage of history, not a superior mind able to conceive the whole, or the philosophical category of

---

[119] Rebenich, p. 91.

[120] Victor Hugo, for example, refused to submit to dictatorship and lived for 19 years in exile, from 1856 to 1870 at Guernsey, in the Channel Islands.

Gesamtheit. In his adversaries he saw *Feinde*, enemies, not *Gegner*, or adversaries who, with their contrasting views contribute to the whole, to a true *intelligence* of what society is.

This judgment on our part does not mean that an intellectual should abstain from political struggles, or should follow the principle of non-participation. It means that an intellectual as a citizen has the duty of taking part, but with the superior consciousness of the role of the others and of their own reasons. A society makes real progress only if all the actors are represented on the political stage. If one moves from the principle that some of them can be excluded, he makes the greatest error one can make, and opens the way up to tyranny.[121]

We will praise in Mommsen the vigorous writer, even on the pages in which he delivers his unfair Cicero judgment. Nevertheless we maintain that he was far from being a real liberal, even if we risk that Rebenich, a writer who shows an undoubted finesse in his work, will style us among the *conservative* enemies of Mommsen.

As for the bibliography in Italian, *Theodor Mommsen e l'Italia*,[122] a collection of essays c/o the Accademia dei Lincei, should be mentioned. It is a useful contribution to a better knowledge of the celebrated German historian and of his relations with his Italian counterparts.

An edition of the *Lettere di Mommsen agli Italiani* is furthermore in preparation on the part of an editorial board, whose seat is in the 'Dipartimento di Diritto Romano' of the Faculty of Law, University of Rome 'La Sapienza', under the editorship of Marco Buonocore. The aim is that of recovering, as much as possible, all the letters that Mommsen wrote to Italian scholars. The first of the letters until now traced goes back to 1845, the last to 1903, the year of Mommsen's death. The book, in at least two volumes, is scheduled for 2017, for the second centenary of the historian's birth. It will be published by the Biblioteca Vaticana in the series *Studi e Testi*, and will clarify the complex problem of the relations between German and Italian worlds of learning in the second half of the nineteenth century.

---

[121] I am here echoing Croce's fundamental concepts, as in the *Manifesto degli Intellettuali Anti-Fascisti*, that is my philosophical and political Bible.

[122] Op. cit.

Cosmology

As for cosmology, as we have demonstrated above, Cicero's works are mostly neglected by historians who write on the eighteenth century, especially on Scottish and British eighteenth century.

Paolo Casini, for example, in his splendid *L'Universo Macchina*[123] ignores Cicero, who isn't mentioned at all in the index of names. This is quite surprising, because Newtonianism, as all the philosophic and scientific systems, was an attainment that did not have an abrupt origin. It needed centuries, if not millennia, of reflection.

Certainly, modern historians can well consider as superficial Middleton's judgment, who wrote, in 1741, that

> "several of the fundamental principles of the modern philosophy which pass for the original discoveries of these later times, are the revival rather of ancient notions maintained by some of the first philosophers of whom we have any notions in history".[124]

And we will not deny the advances in seventeenth and eighteenth-century science, although these advances do not seem to influence the writers on political and philosophical subjects.

Ciceronianism was certainly not at the core of Casini's research, yet it cannot be simply ignored.

Concluding remarks

Middleton's *Life of Cicero* was successful in establishing a reputation for Cicero, while our little volume aims at making the point on Ciceronian studies, in this beginning of the twenty-first century, bringing new light on some particular points, while Cicero's political ethics is progressively recognized.

Today's world is quite different from Mommsen's world, and one could justify the great German historian by writing that his *Römische Geschichte*, after all, was an early work, almost a diversion from the main line of his lasting scholarly achievements in law subjects and in philology. This could be one of the reasons why he did not continue his *Römische Geschichte*, and never wrote 'the fourth

---

[123] Laterza (Bari, 1969).
[124] See above, chapt. 1, Mommsen and Cicero, text coresp. with n. 141.

volume'. A reputation for Cicero and his political and philosophical ideas needs now to be definitively re-established, in this world that doesn't seem to be ready to accept ideas of liberty, of tolerance and of respect for human beings, that should be the common ideal of all of us.

Appendix A: Montesquieu, Abbé Galiani, Henry H. Milman, Moses S. Slaughter on Cicero, Drumann, Mommsen

a) Montesquieu, *Discours sur Cicéron*, first published in 1892 in *Mélanges Inédits*, pp. 3-11. The text is from Montesquieu, *Œuvres Complètes*, édition établie et annotée par Roger Caillois, La Pléiade, (Paris, 1949), vol. I, pp. 93-98.

Mommsen, when publishing his *Römische Geschichte* (1854-56), was unaware of this short work of Montesquieu, in which the author of *The Spirit of the Laws* had said exactly the contrary of what he wrote. Nevertheless Montesquieu adds, in a note, the following words: "J'ai fait ce discours dans ma jeunesse. Il pourra devenir bon, si je lui ôte l'air de panegyrique".

This work is admirable for its profound critical insight, for the extraordinary capacity of the great French writer of addressing the problems raised by historical criticism, and of giving them a substantial response, almost line by line.

The present author fully agrees with the words of Montesquieu.

"Cicéron est, de tous les anciens, celui qui a eu le plus de mérite personnel, et à qui j'aimerois mieux ressembler ; il n'y en a aucun qui ait soutenu de plus beaux et de plus grands caractères, qui ait plus aimé la gloire, qui s'en soit fait une plus solide, et qui y ait été par des routes moins battues.

La lecture de ses ouvrages n'élève pas moins le cœur que l'esprit : son éloquence est toute grande, toute majestueuse, toute héroïque. Il faut le voir triompher de Catilina ; il faut le voir s'élever contre Antoine ; il faut le voir enfin pleurer les déplorables restes d'une liberté mourante. Soit qu'il raconte ses actions, soit qu'il rapporte celle des grands hommes qui ont combattu pour la

République, il s'enivre de sa gloire et de la leur. La hardiesse de ses expressions fait entrer dans la vivacité de ses sentiments. Je sens qu'il m'entraîne dans ses transports et m'enlève dans ses mouvements. Quels portraits que ceux qu'il fait des Brutus, des Cassius, des Catons ! Quel feu, quelle vivacité, quelle rapidité, quel torrent d'éloquence ! Pour moi, je ne sais à qui j'aimerois mieux ressembler, ou au héros, ou au panégyriste.

S'il relève quelquefois ses talents avec trop de faste, il ne fait que m'exprimer ce qu'il m'avoit déjà fait sentir ; il me prévient sur des louanges qui lui sont dues. Je ne suis point fâché d'être averti que ce n'est pas un simple orateur qui parle, mais le libérateur de la patrie et le défenseur de la liberté.

Il ne mérite pas moins le titre de philosophe que d'orateur romain. On peut dire même qu'il s'est plus signalé dans le Lycée que sur la tribune : il est original dans ses livres de philosophie, mais il a plusieurs rivaux de son éloquence.

Il est le premier, chez les Romains, qui ait tiré la philosophie des mains des savants, et l'ait dégagée des embarras d'une langue étrangère. Il la rendit commune à tous les hommes, comme la raison, et, dans les applaudissements qu'il en reçut, les gens de lettres se trouvèrent d'accord avec le peuple. Je ne puis assez admirer la profondeur de ses raisonnements dans un temps où les sages ne se distinguoient que par la bizarrerie de leur vêtement. Je voudrois seulement qu'il fût venu dans un siècle plus éclairé, et qu'il eût pu employer à découvrir des vérités ces heureux talents, qui ne lui ont servi qu'à détruire des erreurs. Il faut avouer qu'il laissa un vide affreux dans la philosophie : il détruisit tout ce qui avoit été imaginé jusqu'alors ; il fallut recommencer, et imaginer de nouveau ; le genre humain rentra, pour ainsi dire, dans l'enfance, et il fut remis aux premiers principes.

Quel plaisir de le voir, dans son livre *De la Nature des Dieux*, faire passer en revue toutes les sectes, confondre tous les philosophes, et marquer chaque préjugé de quelque flétrissure! Tantôt il combat contre ces monstres; tantôt il se joue de la philosophie. Les champions qu'il introduit se détruisent eux-mêmes; celui-là est confondu par celui-ci, qui se trouve battu à son tour. Tous ces systèmes s'évanouissent les uns devant les autres, et il ne reste, dans l'esprit du lecteur, que du mépris pour les philosophes et de l'admiration pour le critique.

Avec quelle satisfaction ne le voit-on pas, dans son livre *De la Divination*, affranchir l'esprit des Romains du joug ridicule des aruspices et des règles de cet art, qui étoit l'opprobre de la théologie païenne, qui fut établi dans le commencement, par la politique des magistrats, chez des peuples grossiers, et affoiblis, par la même politique, lorsqu'ils devinrent plus éclairés.

Tantôt il nous dévoile les charmes de l'amitié et nous en fait sentir tous les délices; tantôt il nous fait voir les avantages d'un âge que la raison éclaire, et qui nous sauve de la violence des passions.

Tantôt, formant nos mœurs et nous montrant l'étendue de nos devoirs, il nous apprend ce que c'est que l'honnête et ce que c'est que l'utile; ce que nous devons à nous-mêmes, ce que nous devons faire en qualité de pères de famille ou en qualité de citoyens.

Ses mœurs étoient plus austères que son esprit. Il se comporta dans son gouvernement de Cilicie avec le désintéressement des Cincinnatus, das Camilles, des Catons. Mais sa vertu, qui n'avoit rien de farouche, ne l'empêchoit point de jouir de la politesse de son siècle. On remarque, dans ses ouvrages de morale, un air de gaieté et un certain contentement d'esprit que les philosophes médiocres ne connoissent point. Il ne donne point de précepts; mais il les fait sentir. Il n'excite pas à la vertu; mais il y attire. Qu'on lise ses ouvrages, et on sera dégoûté pour toujours de Sénèque et de ses semblables, gens plus malades que ceux qu'ils veulent guérir, plus désespérés que ceux qu'ils consolent, plus tyrannisés des passions que ceux qu'ils en veulent affranchir.

Quelques personnes, accoutumées à mesurer tous les héros sur celui de Quinte Curce, se sont fait de Cicéron une idée bien fausse ; ils l'ont regardé, comme un homme foible et timide, et lui ont fait un reproche qu'Antoine, son plus grand ennemi, ne lui a jamais fait. Il évitoit le péril, parce qu'il le connoissoit ; mais il ne le connoissoit plus lorsqu'il ne pouvoit plus l'éviter.

Ce grand homme subordonna toujours toutes ses passions, sa crainte et son courage, à la sagesse et à la raison. J'ose même le dire : il n'y a peut-être point d'homme, chez les Romains, qui ait donné de plus grands examples de force et de courage.

N'est-il pas vrai que déclamer la Seconde Philippique devant Antoine, c'étoit courir à une mort certaine ? C'étoit faire un généreux sacrifice de sa vie en faveur de sa gloire offensée ? Admirons donc le courage et la hardiesse de l'orateur encore plus que son éloquence. Considérons Antoine, le plus puissant d'entre les hommes, Antoine, le maître du monde, Antoine, qui osoit tout et qui pouvoit tout ce qu'il osoit, dans un Sénat qui étoit entouré de ses soldats, et où il étoit plutôt roi que consul ; considérons-le, dis-je, couvert de confusion et d'ignominie, foudroyé, anéanti, obligé d'entendre ce qu'il y a de plus humiliant de la bouche d'un homme à qui il auroit pu ôter mille vies.

Aussi, ce ne fut pas seulement à la tête d'une armée qu'il eut besoin de sa fermeté et de son courage ; les traverses qu'il eut à souffrir, dans des temps si difficiles pour les gens de bien, lui rendirent la mort toujours présente. Tous les ennemis de la république furent les siens : les Verrès, les Clodius, les Catilinas, les Césars, les Antoines, enfin tous les scélérats de Rome lui déclarèrent la guerre.

Il est vrai qu'il y eut des occasions où la force de son esprit sembla l'abandonner : lorsqu'il vit Rome déchirée par tant de factions, il se livra à la douleur, il se laissa abattre, et sa philosophie fut moins forte que son amour pour la République.

Dans cette fameuse guerre qui décida de la destinée de l'Univers, il trembloit pour sa patrie ; il voyoit César approcher avec une armée qui avoit gagné plus de batailles qu'elle n'avoit de légions. Mais quelle fut sa douleur lorsqu'il vit que Pompée abandonnoit l'Italie et laissoit Rome exposée à la fureur des rebelles !

<Après une telle lâcheté, dit-il, je ne puis plus estimer cet homme, qui, bien loin de s'exiler de sa patrie, comme il a fait, devoit mourir sur les murailles de Rome et s'ensevelir sous ses ruines>.

Cicéron, qui étudioit depuis longtemps les projets de César, auroit fait subir à cet ambitieux le destin de Catilina, si sa prudence eût été écoutée : <Si mes conseils avoient été suivis, dit cet orateur à Antoine, la République fleuriroit aujourd'hui, et tu serois dans le néant. Je fus d'avis qu'il ne falloit point continuer à César le gouvernement des Gaules au-delà des cinq ans. Je fus d'avis encore que, pendant qu'il seroit absent, l'on ne devoit point l'admettre à demander le consulat. Si j'avois été assez heureux pour persuader l'un ou l'autre, nous ne serions jamais tombés dans l'abîme où nous sommes aujourd'hui. Mais, lorsque je vis (continue-t-il) que Pompée avoit livré la République à César, quand je m'aperçus quil commençoit trop tard à sentir les maux que j'avois prévus depuis si longtemps, je ne cessai pour lors de parler d'accommodement, et je n'épargnai rien pour réunir les esprits>.

Pompée ayant abandonné l'Italie, Cicéron, qui, comme il le dit lui-même, savoit bien qu'il devoit fuir, mais ignoroit qui il devoit suivre, y resta encore quelque temps. César s'aboucha avec lui et voulut l'obliger, par prières et par menaces, de se ranger de son parti. Mais ce républicain rejeta ses propositions avec autant de mépris que de fierté. Lorsque le parti de la liberté eut été détruit, il se soumit à lui avec tout l'Univers ; il ne fit point une résistance inutile ; il ne fit point comme Caton, qui abandonna lâchement la République avec la vie ; il se réserva pour des temps plus heureux, et il chercha dans la philosophie des consolations que les autres n'avoient trouvées que dans la mort.

Il se retira à Tusculum pour y chercher la liberté, que sa patrie avoit perdue. Ces champs ne furent jamais si glorieusement fertiles ; nous leur devons ces beaux ouvrages qui seront admirés par toutes les sectes et dans toutes les révolutions de la philosophie.

Mais, lorsque les conjurés eurent commis cette grande action qui étonne encore aujourd'hui les tyrans, Cicéron sortit comme du tombeau, et ce soleil, que l'astre de Jules[1] avoit éclipsé, reprit une nouvelle lumière. Brutus, tout couvert de sang et de gloire, montrant au peuple le poignard et la liberté, s'écria : <Cicéron !>. Et, soit qu'il l'appelât à son secours, soit qu'il voulût[2] le féliciter de la liberté qu'il venoit de lui rendre, soit enfin que ce nouveau libérateur de la patrie se déclarât son rival, il fit de lui dans un seul mot le plus magnifique éloge qu'un mortel ait jamais reçu.

Cicéron se joignit aussitôt à Brutus ; les périls ne l'étonnèrent point. César vivoit encore dans le cœur de ses soldats ; Antoine, qui étoit l'héritier de son ambition, tenoit dans ses mains l'autorité consulaire. Tout cela ne l'empêcha point de se déclarer, et, par son autorité et son example, il détermina l'Univers encore incertain s'il devoit regarder Brutus comme un parricide ou comme le libérateur de la patrie.

Mais les libéralités que César avoit faites aux Romains par son testament furent pour elles de nouvelles chaînes.

Antoine harangua ce peuple avare, et, lui montrant la robe sanglante de César, il l'émut si fort qu'il alla mettre

---

[1] Julius Sidus (Montesquieu's note).
[2] Seconde Philippique (Montesquieu's note).

le feu aux maisons des conjurés. Brutus et Cassius, contraints d'abandonner leur ingrate patrie, n'eurent que ce moyen pour se dérober aux insultes d'une populace aussi furieuse qu'aveugle.

Antoine, devenu plus hardi, usurpa dans Rome plus d'autorité que n'avoit fait César même. Il s'empara des deniers publics, vendit les provinces et les magistratures, fit la guerre aux colonies romaines, viola enfin toutes les lois. Fier du succès de son éloquence, il ne redouta plus celle de Cicéron, il déclama contre lui en plein Sénat ; mais il fut bien étonné de trouver encore dans Rome un Romain.

Bientôt après, Octave fit cet infâme traité dans lequel Antoine, pour prix de son amitié, exigea la tête de Cicéron. Jamais guerre ne fut plus funeste à la République que cette indigne réconciliation, où l'on n'immola pour victimes que ceux qui l'avoient si glorieusement défendue.

Le détestable Popilius est justifié ainsi, dans Sénèque, de la mort de Cicéron : que ce crime si odieux étoit le crime d'Antoine, qui l'avoit commandé, non pas celui de Popilius, qui avoit obéi ; que la proscription de Cicéron avoit été de mourir, celle de Popilius de lui ôter la vie ; qu'il n'étoit pas merveilleux qu'il eût été forcé de le tuer, puisque Cicéron, le premier de tous les Romains, avoit été contraint de perdre la tête."[3]

b) Abbé Galiani: Abbé Ferdinando Galiani, the celebrated author of the *Dialogue sur le Commerce des Bleds* (Paris-London, 1770), a favourite in the Parisian salons, on 20 July 1771 wrote a letter from Naples to M.me D'Épinay, that needs to be here transcribed: see *Lettres de l'Abbé Galiani à M.me D'Épinay, in two vols*, (Paris, 1882), vol. 1, pp. 255-258.

The portrait that he gives of Cicero is certainly bizarre, and it aims at flattering M.me D'Épinay. Nevertheless some relevant points should be emphasized, especially where the author maintains that "le parti de l'opposition était un parti d'incrédules; car

---

[3]Septième Controverse (Montesquieu's note).

les évêques (c'est-à-dire les augures, les pontifs, etc.) étaient tous lords et patriciens.... Ainsi Cicéron qui, dans son cœur, penchait du côté de l'opposition, était incrédule à cachette, et n'osait pas le paraître."

"On peut regarder Cicéron comme littérateur, comme philosophe et comme homme d'état. Il a été un des plus grands littérateurs de son temps. Il savait tout ce qu'on savait de son temps, excepté la géometrie et autres sciences de ce genre. Il était médiocre philosophe, car il savait tout ce que les Grecs avaient pensé, et le rendait avec une clarté admirable ; mais ne pensait rien, et n'avait pas la force de rien imaginer. Il eut l'adresse et le bonheur à rendre, en langue latine, les pensées des Grecs, et cela le fit lire et admirer par ses compatriotes ...

Comme homme d'état Cicéron était d'une basse extraction et voulant parvenir, aurait dû se jeter dans le parti de l'opposition, ou de la chambre basse, ou du people, si vous voulez. Cela lui était d'autant plus aisé, que Marius, fondateur de ce parti, était de son pays. Il en fut même tenté : car il débuta par attaquer Sylla, et par se lier d'amitié avec les gens du parti de l'opposition, à la tête desquels, après la mort de Marius, étaient Clodius, Catilina, César. Mais le grand parti avait besoin d'un juresconsulte et d'un savant ; car les grands seigneurs, en général, ne savent ni lire ni écrire. Il sentit donc qu'on aurait plus besoin de lui dans le parti des grands, et qu'il y jouerait un rôle plus brillant. Il s'y jeta, et dès lors on vit un nouveau parvenu mêlé avec les patriciens ...

Cicéron brilla donc à côté de Pompée, etc, toutes fois qu'il était question de choses de jurisprudence ; mais il lui manquait la naissance, les richesses et, surtout n'étant pas homme de guerre, il jouait de ce côté-là un rôle subalterne. D'ailleurs, par inclination naturelle, il aimait le parti de César, et il était fatigué de la morgue des grands, qui lui faisaient sentir souvent la grandeur

des bienfaits dont on l'avait comblé. Il n'était pas pusil-
lanime : il était incertain ...

Pour les vertus de Cicéron on n'en sait rien : il ne gou-
verna jamais. Pour ce qui est de son mérite d'avoir ou-
vert les portes de Rome à la philosophie, il est bon de
dire que le parti de l'opposition était un parti d'incrédules ; car les évêques (c'est-à-dire les augures, les pontifs, etc.) étaient tous lords et patriciens. Ainsi, le parti
de l'opposition attaquait la religion, et Lucrèce avait
écrit son poème avant Cicéron. Ainsi Cicéron qui, dans
son cœur, penchait du côté de l'opposition, était incrédule à cachette, et n'osait pas le paraître. Lorsque le
parti de César triompha, il se montra plus à découvert
sans en rougir. Mais ce n'est pas à lui qu'on doit la
fondation de l'incrédulité païenne, qu'il appelait sophia,
sagesse ; c'est au parti de César. Les applaudissements
que la postérité a donnés à Cicéron, viennent de ce qu'il
suivit le parti contraire à celui que la cruauté des empe-
reurs rendit odieux.

Et voilà assez sur Cicéron".

c) From *The London Quarterly Review*, American Edition, June
1836, No CXII, art. IV, pp. 182-200. The journal was at that time
(1826-1853) edited by John Gibson Lockhart. The review of the
book by Drumann is credited to the Reverend Henry Hart Milman
(1791-1868): see the John Murray Archive at the National Library
of Scotland, ref. acc. 13236/55.

Milman was professor of poetry at Oxford, the author of the
History of Latin Christianity, of a *Life of Gibbon*, and the editor of
Gibbon's *Decline and Fall of the Roman Empire*.

The six Drumann volumes were published in 1834-1844, therefore
Milman immediately perceived the anti-Ciceronian stance of the
work, although only the last two books are on Cicero, in book V
pages 218-697, and the whole of vol. VI. Typical is the fact that
this review appeared in 'liberal' England, in contrast to the political
viewpoint then prevailing in Prussia. On the same ideological line
we will find the subsequent reviews and works in Anglo-American
historiography.

"Drumann ... an avowed and ardent admirer of monarchy ... No subject of the present King of Prussia, he asserts, doubts that monarchy is the best form of government.

His conclusion is that Rome only reaches the heaven of monarchy as a refuge from the horrors and exhaustion of the civil war- The event could not be otherwise.

Nevertheless Cicero maintained his influence, his commanding powers in civil life.

Drumann does not do justice to the character of Cicero-the embarrassment of Cicero's position, his necessarily less commanding attitude as a man of peace, among rivals at the head of their respective armies, might have excused more vacillations and uncertainty that can fairly be charged on the great orator- how nobly Cicero contrasts with the rest of his contemporaries ...

While others imported the wealth of the East to corrupt the suffrages of the people, its sensual luxuries to debase their morals, even its superstitions to infect their religion, Cicero transplanted into Rome the wisdom, the taste, the philosophy of Greece. The treasures which he accumulated from captive provinces were the books, the works of art, the teachers of oratory or of philosophy. If he did not neglect his golden opportunities in the more barbarous provinces of Asia which he administered, he laid out his wealth for the intellectual, and therefore, at such a period, the moral advancement of his country. Who was ever so great in the double character of the statesman and the man of letters?

The man of letters never repressed the energy, or enthralled the activity of the statesman; the public man, excepting in the orations, which, in fact, are the literature of the public man, and the letters which lay open his public sentiments and conduct as well as his private feelings, the busy political leader does not disturb the serene dignity of the man of letters. Notwithstanding the sarcasm perpetuated by Juvenal, we could quote passages to show that Cicero might have been a poet of a very high order; as a Roman orator standing alone; as a philosopher, if neither very profound nor original, condensing all the wisdom of the different schools of Greece in his perspicuous and vivid Latinity; at the same time he was the main support of the senatorial or constitutional party against all opponents- against Catiline, Clodius, Antony.

The following sentence is surely exempt from the charge of unwarrantable egoism: <*Me nec reipublicae, nec amicis unquam defuisse, et tamen omni genere monumentorum meorum perfecisse operis subsecivis, ut meae vigiliae, meaeque literae, ut juventuti utilitatis, et nomini romano laudis aliquid adferrent>.*[4]

These general and desultory remarks, we are aware, by no means do justice to a work, the chief excellence of which consists in its minute and laborious accuracy of detail ...

Antony ... first succeeded almost in convincing Cicero of his patriotic designs ... Cicero and Antony were now the real heads of their respective parties; Cicero of that of the constitution, of the aristocracy, of the government by civil authority, Antony of that of monarchy, however disguised, of the people ... Could Cicero have done more in his position, and with the means at his command? He was weighed down in first place by the imbecillity and misconduct of the conspirators ... (as for the *Philippics*) Cicero had a prophetic consciousness of the peril, though he did not or could not shrink from the responsibility of his position ... With Cicero ... fell the liberties of Rome."

---

[4] "I ... while never failing in my obligations to the Republic and to my friends, have employed my spare hours in producing works in a variety of genres that will preserve my name, in order that what I have written in the watches of the night may be of some profit to your young people and bring some credit to the Roman name", *Phil.* 2.20, trans. Loeb Classical Library.

d) Moses Stephen Slaughter, *Cicero and his Critics*, *The Classical Journal*, XVIII, 1921-22, pp. 120-131, ibid., passim.

Slaughter (1860-1923) was professor of Latin at the University of Wisconsin. In the world catalogue this book appears under his name: *Memoirs, Correspondence, Memorabilia, and printed matter, relating to relief activities of the Red Cross in Italy at the end of World War I and the occupation of Fiume by the forces of G. D'Annunzio. Includes papers of Gertrude Slaughter, wife of M. S. Slaughter.* It has not been possible to consult the book. The Slaughter papers are the property of the University of Wisconsin.

"... It is quite beside the mark to protest that Cicero's philosophical works are of no real or independent value. Mommsen says, <with equal peevishness and precipitation Cicero composed in a couple of months an entire philosophical library>. It is true that in rapid succession, in a short space of time -about two years- at the most distressful period of his life and of that of the Republic, Cicero published a number of treatises on philosophical subjects, mainly ethical, largely adaptations from the Greek. They are not without merit in themselves though they may count for little in the history of philosophy by the side of their Greek equivalents. To the Church Fathers of the fourth and succeeding centuries and to the occasional layman, Cicero was one of those who,

quasi cursores vitae lampada tradunt (Lucr. II:79)-
like runners hand on the lamp of life.

To this audience all this has long been familiar ...

Until the time of Niebuhr, who in the early part of the 19th century wrote the first modern history of Rome, Cicero continued to hold this high place in the councils of humanism. Niebuhr says that "Cicero followed truth in every way, and in his doing so we recognize the discord of his mind; he was in contradiction with himself". Then more kindly Niebuhr adds "I love Cicero as if I had known him and I judge of him as I would of a near relation who had committed a folly". This patronizing compliment is the last kind word we hear for Cicero from professors at the universities of Germany who write on Roman history. With the exception, perhaps, of Ihne, they all seem to suffer from literary adenoids, if an allusion to the *emunctae naris* of Horace (Sat. I, IV, 8) be not too remote.

Cicero's 'folly' grew to a crime in the minds of Mommsen, Drumann's unscrupulous successor in the business of character defamation. The crime for which Cicero is maligned by these anything but self-effacing critics is an incurable faith in a free state.[5]

During the four hundred years from Petrarch to Niebuhr men the world over sat at Cicero's feet and learned from him the secret of practical, literary and spiritual life. Like Virgil he was
Duplici circumdatus aestu carminis et rerum (Manilius)
    Surrounded by a double tide of life and letters.

To him men turned for those studies which he himself glorifies in the oration for Archias, his old teacher:
Haec studia adulescentiam alunt, senectutem oblectant, secundas res ornant, adversis perfugium ac solacium praebent, delectant domi, non impediunt foris, pernoctant nobiscum, peregrinantur, rusticantur.

Such studies nourish youth, delight old age, adorn success, furnish a refuge and a solace in adversity, at home they charm, abroad they do not embarrass, in the night seasons they are with us, they travel with us to the country and to foreign lands.

This is humanism as a grace of life, ministering to the pleasure and polish of social intercourse, as well as a discipline "which aims at drawing out all the mental and moral faculties of man" (Jebb.) ... Mommsen calls Cicero a mystery of human nature; language and the effect of language on the mind a problem which cannot be solved. The power which language exercises was,

---

[5]Italics mine.

he says, in Cicero's case transferred to the unworthy vessel. Custom and the schoolmaster completed what the power of language had begun. Cicero became the supreme stylist and the creator of the modern classical Latin prose. Naturally Mommsen finds nothing to praise in this phase of Cicero's influence. His eloquence lacks fire, his speeches lack clearness and articulate division, and his language is deficient in precision and chasteness. His dialogue is not as good as the Greek -which is doubtless true- but more to the point Mommsen's appraisement is not as good as Lessing's. Of Cicero's correspondence Mommsen says with a sneer that people are in the habit of calling it interesting and clever. To such and to all admirers of Cicero's writings he gives his sovereign, imperious and peremptory advice, "to observe in literary matters a becoming silence"; -so far have we come from Quintilian, a second-century Roman critic whose frequently quoted opinion is. "The pleasure a man takes in Cicero is the standard by which he may judge his own intellectual culture".

Wherever republican institutions have flourished or men have struggled to attain free government, there Cicero has been a quickening influence.[6] The Church Fathers -the 'bitter enders' at least- were in the habit of saying that only Lethe could take away the influence of Cicero ...

By virtue of his wide range of interests, his encyclopedic information in history, literature, and law, his remarkable success in the affairs of state, Cicero has a clear right to the attention of men. His was no single-track mind. The indecision and vacillation of which he is accused and is guilty are due to the necessity he was under of seeing both sides of the question; the arguments pro and con alternately appealed to him and troubled him and rendered decision difficult. He had not the practical aptitude for politics which, according to Mommsen,

---

[6]Italics mine.

made Julius Caesar a 'perfect man'. Caesar possessed, Mommsen says, practical aptitude as a citizen in perfection. Caesar was a thorough-going realist and this made of him the consummate statesman. Caesar's cool sobriety, marvelous serenity, his rationalism, appeal strongly to Mommsen's imperialistic mind. From such a critic we need expect no grace for Cicero either in literature or state-craft.

As most of you know and many of you have said, the modern attitude towards Cicero of harsh criticism and deliberate undervaluation is largely due to the influence and wide circulation and acceptance of Mommsen's *History of Rome. Men living under free institutions have accepted without question the judgment of an arch supporter of a benevolent despotism, who in making a demigod of Julius Caesar, the real founder of the Roman Empire, found it impossible to see any virtue in Cicero, the advocate of a free state, the believer in an ideal republic, and the author of a programme to establish such a state.*[7]

That his cause was a 'lost cause', a forlorn hope, endeared it all the more to Cicero, a man of humanistic ideals, and it should recommend him to the favourable consideration of all men of like tastes and a similar philosophy of life. Cicero says he "*mourned for the commonwealth longer and more bitterly than ever a mother mourned for her only son*"[8] (Ad Att. IX, 20, 3).

Cicero repeatedly gave up politics for which he was not primarily intended by nature, repeatedly returned to the life best suited to his natural disposition, to letters and the studies that he loved. But he could not live a detached life and he was repeatedly drawn back into the political arena, gifted lawyer that he was, at the solicitation of his friends, or because of the ever-returning hope that he might see at last established at Rome that

---

[7]Italics mine.
[8]Italics mine.

form of government in which he believed and for which he prayed. He essayed a difficult and dangerous part in the last years of the Republic. He found in Pompey a broken reed; he saw in Caesar a gracious and attractive personality, with a genius for statemanship but feared in him this genius because it was misdirected and aimed at personal gains. "It is impossible for me", he says of Caesar, "to be other than a friend to one who deserves well of his country".

"Cicero thought the Republic had swooned under Caesar's blow. He did not realize that it was killed" (Strachan-Davidson). He begs Caesar to "have regard for the judge who will come in ages after", for the judgment of posterity, telling him that it was his chief duty to subdue his personal inclinations, to master his angry feelings, to be moderate in victory, and assuring him that a man's chief glory is to "be remembered for great services done to one's friends, one's country, and to mankind". To Caesar's credit it must be said that Cicero's appeals to his clemency and humanity in individual cases were never denied. On the larger questions of state Cicero made little impression upon Caesar's plans for personal control. Once more he retired to his books and the quiet of his country estates. It was only after the assassination of Caesar that Cicero felt all shackles fall from him and entered the forum for the last time for the final contest with Antony. He could have had no illusion as to the probable outcome of this struggle but he made good use of his freedom and followed what he conceived to be his duty to his death. In no period of his career does Cicero appear to better advantage than in this last encounter with Antony and the men who sought only self-aggrandizement and whose final success meant the overthrow of Cicero's dream of a free state and the permanent establishment of a government with Augustus Caesar in sole control ...

Cicero died like a Roman, and by so doing atoned for many littlenesses: vanity, conceit, ultra-sensitiveness, exhibitions of physical timidity, bordering on physical cowardice, if atonement is asked for such things from one whose purity of life and high moral standards in all personal dealings combine to make of him a shining exception among the men of his day.

Cicero failed in the one consuming desire of his life, to see a free state established at Rome and to be not its ruler but a participator in its benefits and a sharer in the glory of its success. He had many gifts of the statesman, but Mommsen says he lacked courage and "on those who lack courage, the gods lavish every favour and every gift in vain". His hero, Caesar, had courage and destroyed the Republic.

Commenting on Caesar's successful usurpation of the powers of the Republic, Mommsen in one of his choicest sentences, which may be of interest to you as throwing light on the animus with which he wrote the story of the fall of the Roman Republic, says: "Not until the dragon seed of North America ripens will the world have again similar fruits to reap". To depart from history to indulge in prophecy is not the only mistake Mommsen makes in this astonishing sentence.

The restoration of the Republic of the Scipios was Cicero's solution for the world's ills and to this he clung, perhaps mistakenly, until the end. He has left us a picture of his ideal republic in his *De Republica*, of which unfortunately only fragments have come down to us. The world has united in praising this document, but Mommsen calls it a "singular mongrel compound of history and philosophy, which carries out the idea that the existing constitution of Rome is substantially the ideal state organization sought for by the philosophers, an idea as unphilosophical as it is unhistorical".

Mommsen is quite incapable by birth, nature, and training of understanding the situation in Rome in Cicero's day seen from any idealistic point of view. The idea of a free state enrages him. He worships Caesar and de-

fends the despotism established by him, the revised divine right despotism brought into Europe from the Orient by Alexander of Macedon. The very name Caesar intrigues the mind of Mommsen and betrays him into making this self-revealing statement which fortunately now calls for a slight revision: "the peoples to whom the world belongs still at the present day designate the highest of their rulers by the name Caesar"[9].

After such a pronouncement have we not the right to demand of the German historians that they set a less biased man to the task of rewriting the story of the Roman Republic, one who knows neither *ira* nor *studio*, and may we not expect at his hands a fairer treatment of the man whose unpardonable sin was a belief in free institutions? Having all his utterances in mind may it not be pertinent to bid Mommsen and his kind "to observe in matters historical a becoming silence"? ...

Livy once said: "To praise Cicero as he deserves we ought to have another Cicero". I have not sought to supply the deficiency but I have sought to warn you once more against the unwarranted defamation of his character and the belittling of his achievements and ideals at the hands of historians too biased to appreciate either.

To know Cicero well is to live a large life in the midst of the stirring events of one of the world's greatest eras, an era with which our own has many points in common. He, too, had to live and work with plutocratic republicans whose minds were solid and with radical republicans whose minds were fluid ... That he lost the battle does not detract from our interest in the struggle. That he preached better than he at all times practiced does not rule him out of our class. That in the main Cicero held to his high ideals cannot truthfully be gain said. His letters discover to us his human weaknesses -it's a hard test- but they also mirror for us a personality affable, lovable, affectionate. He had a pretty wit,

---

[9]Italics mine.

scintillating at times, at times even scurrilous, and a penetrating power of phrase that made of him a terror to his opponents or enemies but a constant delight to his friends. He could use his vocabulary with astonishing and refreshing freedom. Caesar appreciated this gift of Cicero's and never failed to ask for Cicero's latest bon mot. These were the times when of Caesar's despotism it could be said that it was a despotism "tempered by epigrams" -and the best were Cicero's.

Cicero was many, if not myriad, minded. His temperament was mercurial. His life was free from Rome's worst sins -sins to which Caesar openly and unblushingly surrendered. His high-mindedness has never been impeached, and in his character as wise councillor he has been the friendly companion of many noble souls from his to our own time. His writings need no expurgation on the score of morals. Very fittingly has he been given a large place in the discipline of humanism, which seeks to prepare for a life of action by acquaintance with men of thought and action in all times. In literature and philosophy Cicero sought and found the guide of life. When fate dealt him her hardest blow, he found in Plato and in the Greek poets the only consolation a pagan could know. When the state failed him and his home was bereft of his daughter Tullia, he sought in his books, the constant friends and companions of his life, that wisdom and comfort which he himself has so generously handed on to generations of men coming after. His appeal to the judgment of posterity has not been in vain. Those men who have known him best have found in him what the poet's phrase so adequately expresses, "Rome's least mortal mind".[10]

---

[10] "the Roman friend of Rome's least mortal mind/ the friend of Tully", Lord Byron, *Child Harold's Pilgrimage*, Canto IV.44.

Appendix B: Michele Messina's defence of Cicero [1]

MICHELE MESSINA [2]
Giudizi d'illustri scrittori latini italiani e stranieri sul
libro intitolato:
Apologia di Cicerone contro Teodoro Mommsen.

*** ***

Stefano Viglione was an ecclesiastic and Latin scholar, teacher of rhetoric in the seminary of Aversa (Naples), and author of educational writings.

*** ***

Eruditissimo viro

MICHAELI MESSINAE

STEPHANUS VIGLIONIUS

s.p.d. [3]

De egregio et plane aureo illo opere, quod pro Marco Tullio Cicerone adversus Theodorum Mommsenium confecisti, quodque intra vertentem superioris anni Decembrem mensem ad me pro tua benignitate mittendum curasti, vix dici potest, mi carissime Michael, quantum

---

[1] In this section is transcribed the last part of the *Apologia di Cicerone contro Teodoro Mommsen*, Tipografia Eugenio, (Napoli, 1878; second edition 1882), by Michele Messina, with an appendix containing the letters that are here reprinted.

[2] Michele Messina was the author of *La Letteratura Latina in Italia nel secolo XIX* (Napoli, tipografia Giannini, 1876), from which B. Croce, in 'Il giudizio del Mommsen su Cicerone', *Varietà*, op. cit., p. 4, states that he was born in 1850, that at the age of 12 he lived in the house of an uncle of his, a deacon in Moliterno, Southern Italy. Croce adds that there is no further news about his life, and that probably he died in his youth.

[3] *Salutem plurimam dicit*, i.e. sends very warm greetings.

te amem: tantopere me oblectavit! tantam ex-eo lae-
titiae voluptatem cepi! Nihil enim eruditius, nihil ele-
gantius, nihil suavius, nihil magis industria elaboratum,
nihil magis diligentia expolitum atque absolutum. Adeo
singula et universa praeclare disserta, dilucide et per-
spicue tractata, omni ex parte strenue propugnata, af-
fabre illustrata nitescunt. Tu hominum elegantissime,
Michael, totum totum, quantus est, peracris ingenii tui
cultello Marcum Tullium Ciceronem ad pluteum[4] evi-
scerasti, et in particulas concisum ante mercurialium[5]
virorum minime lippientes oculos propius intuendum,
et curiosius perspiciendum in medium attulisti. Atque
inde factum est ut naviter et miro prorsus artificio de eadem fidelia duos parietes dealbaris,[6] turpissimam nimi-
rum insanientis atque impudentissimi hominis personam
Theodoro Mommsenio imposuisti; in quem profecto, ut
equidem censeo, optime quadrant pervulgata illa prover-
bia, asinus ad lyram, sus Minervam:[7] proh scelus! quem
usque ab ultima antiquitate uti absolutissimum recte
sentiendi et scribendi exemplar, universae litterariae rei
benemerentissimum parentem, totius latinitatis uberri-
mum ac lucidissimum fontem, in politicis πολιτικώτατον[8]
et rerum publicarum scientissimum praeceptorem et gu-
bernatorem sapientissimi ac doctissimi quique homines
per omnes deinceps aetates dilexerunt, admirati sunt,
omni laudum ac praeconiorum ratione exornandum ex-

---

[4] Apparently used with the meaning of 'with the bistoury on the anatomical table'.

[5] Specially favoured by Mercury, i.e. the poets: see Horace, *Carmina*, II, 17, 29 ff. In ibid., II, 7, 13-16. Horace says that he was protected by Mercurius.

[6] After Cicero, *Letters to Friends*, Loeb Classical Library, Letter 264, 2, (*Familiares*, 7, 29, 2): "Sine eum errare et putare me virum bonum esse, nec solere duo parietes de eadem fidelia dealbare", "Leave him to his delusion. Let him go on thinking me a man of honour, not one to cover two walls from the same pot of whitewash" (i.e., serving two masters at the same time); trans. Loeb Classical Library.

[7] "Nam etsi non sus Minervam, ut aiunt, tamen inepte quisquis Minervam docet", "even if it is not a case of the proverbial pig teaching Minerva, anyway whoever teaches Minerva is doing a silly thing", Cicero, *Acad.*, I, v, 18; trans. Loeb Classical Library.

[8] Befitting a statesman.

istimarunt, de hoc eodem Marco Tullio Cicerone unum Theodorum Mommsenium tam praepostere, tam mendose, tam foede et sentire, et iudicare, et velut ex Apollinis tripode[9] pronuntiare, ut nihil vilius atque vulgarius debeat aestimari? Tantumne audaciae? Tantumne impudentiae? Tu contra, studiosorum optime, Messina, praesenti tuo apologetico libro nova et pulcherrima illustrioris gloriae corona oratorum cuiuscumque saeculi lumen et ornamentum, imo principem longe praestantissimum ex merito decorasti, atque illi momunentum aere perennius[10] permansurum exegisti.

Ad haec vetustissimum decus Italiae, quae ingeniosorum hominum et hellenicis ac latinis litteris optimorum scriptorum felicissima parens, et beatissima nutrix per omnem tempestatem extitit, multo pluris amplificasti, et clariorem famae nobilitatem eidem conciliasti; ex quo illud quoque sequatur necesse est, ut nostrae studiosae iuventuti (quo sane ardenti desiderio et numquam satis laudanda cupiditate multis abhinc annis distineri mihi alias significasti) ad latinas litteras maiori cum sedulitate et contentione excolendas stimuli acriores subiiciantur; et dudum obdormiscentes latinae litterae tandem aliquando excitentur, novamque lucem ac vitam nanciscantur. Ad extremum multo gratissima ac iucundissima haec tua lucubratio mihi accidit, tum quod a puero non sine magna animi atque ingenii mei utilitate Marco Tullio Ciceroni attente lectitando eiusque pretiosioribus gemmis colligendis cupidissime assuevi, tum quod in Castri Regali Hospitio[11] Marcii Tullii Ciceronis honestissimo nomine, quamquam immerito, fuerim donatus. Hic omnibus de causis maximopere, mi carissime Michael, et Italiae universae, et studiosorum omnium, et imprimis meo ipsius nomine gratulor et exhortor, ut aliis

---

[9] The oracles used to come from the tripod of Apollon.

[10] Horace, *Carmina*, III, 30, 1, "I raised up a monument more durable than bronze". Translation by author.

[11] Unclear, but he seems to refer to a laurel crown that he received at Castro Reale, in Eastern Sicily, in a building that he calls Hospitium, i.e. hospice.

atque aliis huius generis lucubrationibus latinas litteras
fovere non desistas, praecipue in rebus latinis, plumbea
hac nostra aetate praeclarum tibi nomen, novam Italiae amplitudinem, tibi amicissimo Stephano Viglionio
ingentem voluptatem allaturus. Vale

Datum Fractae Maioris Kal. Februarii A. Rep. Sal.[12]
MDCCCLXXIX

*[To the most accomplished man Michele Messina, Ste-
fano Viglione with very warm greetings.*

*Concerning that admirable and clearly splendid work
that you did in defence of Marcus Tullius Cicero against
Theodor Mommsen, and that you kindly took care to
send me early in December last year, I can hardly express, my dearest Michael, how much I like what you
said: to such a degree that it delighted me! So much
pleasure and happiness I got from it! Nothing is in fact
more learned, nothing more elegant, nothing more pleasant, nothing elaborated with more diligence, nothing is
as polished and completed with such care. To such a
high degree both the particulars and the whole are so
clearly arranged, dealt with, with a clearness and lucid-
ity, that from any side they shine as vigorously defended
and skillfully enlightened.*

*You, the most elegant of the men, Michael, with the
knife of your very acute talent have disemboweled Marcus Tullius Cicero with the bistoury on the anatomical
table, and brought it to the centre, cut to pieces, before
the eyes of men protected by Mercurius, suffering by no
means from inflammation of the eyes, that they could
examine closer and investigate more carefully. And it
happened that with energy and with a remarkable skill
you moved straight ahead and from the same bucket you
whitewashed two walls, you inflicted on Theodor Mommsen no doubt a shameful mask; and having been success-*

---

[12] *A reperta salute*, literally 'since when we recovered the salvation', i.e. post
Christum natum.

ful against him, as I for my part believe what you did most aptly fits the well known proverb, asinus ad lyram, sus Minervam: such a villainy!

That man whom the wisest men in all antiquity judged as the most perfect example of the right discernment and writing, as the most deserving father of universal literature, the most copious and shining source of all latinity, in politics a befitting statesman and most knowledgeable authority and helmsman of the state, and having judged that he deserved to be adorned with any praise and encomium, this same Marcus Tullius Cicero of whom only Theodor Mommsen could consider and judge to such a degree out of the proper order, to be full of faults, repugnant, and like from the tripod of Apollon, such that nothing must be considered more contemptible and vulgar?

Why so much boldness? Why so much effrontery? By contrast you, Messina, the best of the scholars, with your apologetic book you have adorned for its merits with a new and very beautiful crown of more shining glory, the light and ornament of the orators of any century, even better the master quite the most outstanding, and raised for him a monument that will last longer than bronze.

Much more, you have added to this very old glory of Italy, which is the most blessed parent of the men of talent and of writers in Greek and Latin letters, and which has been a conspicuous and very fortunate nurse through the ages, and brought it the nobility of a more illustrious fame. From this will necessarily come in due course, that to our studious youth (that you told me, years ago, on another occasion, are distracted from that burning passion and from that desire that we will never praise enough) further spurs are suggested towards cultivating Latin letters with more assiduity and exertion; and when this happens Latin letters, that are now asleep, at last will eventually wake up, acquiring new light and life.

To conclude, this work of you struck me as delightful and was received with deserving gratitude, because, since when I was a boy, with so much advantage for my soul and talent I took the habit of reading Marcus Tullius Cicero carefully, ardently collecting with much cupidity his most precious jewels, so much so that, although undeserving, I received the price, in the hospice of Castro Reale, in the very honourable name of Marcus Tullius Cicero.

For these reasons, in particular, my dearest Michael, I congratulate you and exhort you in the name of the whole of Italy, of all the scholars, and first of all in my name, in order that you do not cease to support Latin letters with more and more works of this kind, above all in Latin matters, and in this leaden age will bring you a splendid name, a new eminence to Italy, and a very great happiness to you, and to Stephan Viglione that great friend of yours. Farewell.[13]]

Frattamaggiore, 1 February, in the year of Our Lord, 1879

*** ***

Ornatissimo, Eruditoque

MICHAELI MESSINA

JOSEPH M.a DE ISANTI s.d.[14]

Quam tibi dederam, mi Michael, iam libero fidem. Opus tuum, quo Marcum Tullium Ciceronem adversus Mommsenii criminationes purgas, perlegi diligenter. Quod quidem opus utrum argumentorum pondus, an elocutionis splendor plus commendet, haud facile dictu fuerit. Nam et rationes tanti momenti sunt ut adversarium deiciant, atque prosternant; si vero tota eloquendi et elegantia,

---

[13]Translation by author.
[14]*Salutem dicit*, i.e. sends his greetings.

et copia spectetur, mihi quidem, cum legerem, quodammodo voluptate abripi videbar. Nihil dicam, qua undique scatet, praeclara eruditione. Quapropter gratulor tibi, quod tam bene de re litteraria mereri cures, quum oratorum omnium maximum, et philosophorum facile principem ab insani hominis calumniis vindicaveris. Quis enim literatorum non gravate ferat, eum, qui a Catullo[15] optimus omnium patronus laudatur, a Divo Hieronimo[16] eloquii flumen praedicatur, a Fabio Quintiliano[17] omnium eminentissimus, atque unicum orandi specimen appellatur, ab homine cerebroso tam male multari? Tibi igitur omnes boni, qui de Marco Tullio ea sentiunt, quae sentire debent, gratiam habendam esse maximam judicabunt. Sed de his hactenus. Reliquum est, ut me tibi purgatum velim, si diem ex die duxerim ad te scribendi. Nam praeterquam quod plurimis occupationibus distineor, tum non tam prospera usus sum valetudine. Malui de tua bonitate confidere. Vale.

Dab. Puteolis V Kal. Martias MDCCCLXXIX

[My dear Michael, I am now accomplishing the promise that I had given you. I have read attentively your work, in which you clear Marcus Tullius Cicero of the charges of Mommsen. It is not easy to say what commends more the work, whether the weight of the reasoning or the splendor of the expression. For, the reasoning is of such a force that it throws the adversary to the ground, and

---

[15] "Disertissime Romuli nepotum/ Marce Tulli/...... gratias tibi maximas Catullus/ agit pessimus omnium poeta/ quanto tu optimus omnium patronus", "Oh Marcus Tullius, the most eloquent of the grandsons of Romulus, expresses his own gratitude to you Catullus, the worst of all the poets, as much as you are the best of the advocates" (Catullus, 49). Translation by author.

[16] "ex flumine tulliano eloquentiae", "from the Tullian river of eloquence", Hieronimus, *Epist.* 36.14. Translation by author.

[17] "in omnibus quae in quoque laudantur eminentissimus", "The most distinguished in all the qualities that are praised in any orator", Quintilian, *Inst.*, 12,10,12; "unicum apud nos specimen orandi", "our sole model of eloquence", Quintilian, ibid, 3,1,20. Translation by author.

causes him to fall; no doubt, looking at the elegance of the expression and at its abundance, while reading I felt as if I were being uplifted by a particular pleasure. I will say nothing as regards your brilliant erudition, that gushes from all sides. Therefore I congratulate you with regard to the fact that you merit so well as for literature, and at the same time you defend the greatest of all the orators and the most distinguished philosopher, from the calumny of that crazy man. And, in fact, who, of the literati, will not but reluctantly tolerate that the man, who by Catullus is praised as the best of the patrons, by the Divus Hieronymus is called a river of eloquence, by Fabius Quintilianus the most eminent of all, and the sole embodiment of the art of eloquence, so heinously is now punished by an enraged man? All the good men, who feel about Marcus Tullius the feelings that they must feel, will judge that they must have the greatest gratitude for you. But enough of this. It remains that I must apologize if I have delayed my answer day after day. For, apart from the most numerous occupations from which I am kept apart, I don't enjoy good health. I have preferred to rely upon your benevolence. Farewell.][18]

Pozzuoli, 25 February 1879

*** ***

Tommaso Vallauri (1801-1897, classical philologist, was Senator of the Kingdom. He was the author of a 'Vocabolario italiano-latino e latino-italiano'.

Vallauri writes that Mommsen returns with insults the kindness that the Italians show towards him, speaks of his 'imprudent ignorance', and adds that some professors of the Italian universities, following the fashion of the Germans, praise both Mommsen and Ritschl and the other 'hyperborean philologists', who spoil the classics with their own utopias and with lessons rummaged in codices 'incorrect and lacking any authority'.

---

[18]Translation by author.

Di Torino, il 25 marzo 1879

Chiarissimo signore,

Non v'ha dubbio, che V. S. ha dato un carpiccio dei buoni al Mommsen, il quale rimerita coi vituperi la cortesia che gli usano gl'Italiani, quando egli viene tra noi a copiare le epigrafi antiche. E' bisogna sempre biasimare l'imprudente ignoranza del Mommsen, e sferzare di santa ragione certi intedescati professori delle Università italiane, che insegnano la storia colle teorie del professore berlinese, e non rifinano di lodare lui e il Ritschl[19] e gli altri filologi iperborei, che ci guastano i classici colle loro utopie e con un subbisso di varie lezioni, razzolate in codici scorretti e senza autorità.

Haec raptim. Vale.[20]

Il Suo dev.mo Tommaso Vallauri

*** ***

Michele Ferrucci (1801-1881), professor in Geneva and Pisa, contributed to the rewriting of the *Latin Lexicon* of Egidio Forcellini and edited the rewriting of the *Lexicon Epigraphicum* of S.A. Morcelli.

Ferrucci praises Michele Messina, and adds that he believes that 'very few people, both in Italy and abroad, share the opinions of Mommsen'.

---

[19] Friedrich Wilhelm Ritschl (1806-1876), best known as editor of the comedies of Plautus. His *Priscae Latinitatis Monumenta Epigraphica* form an introductory volume to the *Berlin Corpus Inscriptionum Latinarum*.

He was the teacher and tutor, among others, of the philosopher Friedrich Nietzsche.

[20] *These few things hastily. Greetings.* Translation by author.

Pisa 15 Dicembre 1878

Mio caro e riverito Signore,

Dal biglietto di visita,[21] che Le inviai giorni orsono, in segno di ringraziamento, ella avrà potuto comprendere quanto io gradii il dono da lei fattomi del Suo bel libro. Ora, rispondendo alla Sua gentilissima lettera del dì 11, voglio rinnovarLe i miei affettuosi ringraziamenti per tanta Sua benevolenza verso di me, e assicurarLa che la sua apologia pel nostro divino Cicerone mi è assai piaciuta, quantunque creda siano pochissimi e qui in Italia e fuori coloro che la pensino come il Mommsen. In ogni modo, io mi congratulo sinceramente con Lei, che continua a difendere i buoni studi, difendendo il più grande degli oratori latini. Da bravo, non lasci la magnanima Sua impresa: riverisca in mio nome l'erudito Suo zio[22] e conservi sempre la Sua preziosa benevolenza al Suo

dev.mo obb.mo servitore

Michele Ferrucci

*** ***

Mauro Ricci (1826-1900), writer, the author of *Prose sacre, morali e filosofiche*, of *L'allegra filologia di Frate Possedonio da Peretola*, which was reprinted in several editions, and of a translation of Homer's *Iliad* into the Florentine vernacular.

Ricci praises Michele Messina, and just observes that he should have mentioned Saint Thomas, who often has recourse to the authority of the Arpinas.

---

[21] A pié di questo sono scritte le parole: pulchre, bene, recte (note by Michele Messina).

[22] See above n. 2.

Firenze, 16 dicembre 1878

Gentilissimo signor Messina,

Uno stupendo argomento Ella ha preso a trattare, e una buona azione ha fatta assumendosi la difesa di Cicerone, e con lui di tanta parte del decoro italiano. Perciò Le è venuto scritto un libro assai interessante. Non occorre che io Le dica se convengo delle Sue ragioni; anzi Le aggiungo che avrei anche voluto veder citato S. Tommaso, il quale spessissimo si fa forte dell'autorità dell'Arpinate. Posto accanto al presente volume quell'altro di Lei,[23] Le debbo dire che Ella progredisce non camminando, ma correndo. Di tutto mi rallegro, e La prego di credermi

Suo devotissimo

Mauro Ricci

*** ***

Giulio Carcano (1812-1884), writer, Senator of the Kingdom, translator of the works of Shakespeare (in 12 volumes, Milano, 1875-82).

Carcano writes that, in reading Mommsen's *Roman History*, he is 'disgusted' by the 'restless longing for criticizing not only Cicero, but also the other greatest examples of Roman glory, always with the ambitious aim of making people believe that everything that is beautiful and good of the ancient times comes from Germany. But the history of human thought cannot be so easily undone'.

Milano, 14 giugno 1879

Chiarissimo Signore,

Nel renderLe, e senza indugio, molte e sincere grazie della cortese Sua lettera e del pregevole volume apologetico di Cicerone, rivendicato contro il Mommsen, mi è caro di unirmi con Lei nel giudizio su quel grande Latino, e nel nobile pensiero. Le dirò che anch'io, leggendo, or fa molti anni, quello storico arditissimo, mi sentìi più di una

---

[23] *La Letteratura Latina in Italia nel Secolo XIX*, see above, n. 2.

volta disgustato di quella sua smania inquieta di gittare a terra non solo Cicerone, ma le altre più grandi imagini della gloria romana, sempre con l'intento ambizioso di far credere che tutto il bello e il buono del tempo antico vien di Germania. Ma la storia del pensiero umano non può essere così facilmente disfatta.

Leggerò con attenta considerazione questo Suo studio letterario, del quale non potei fare per ora che una rapida scorsa; e intanto non posso che farLe animo a continuare nell'opera buona di conservar le tradizioni di quella età classica, pur troppo non curata e disprezzata al tempo nostro.

Mi creda con molta stima

Suo devotissimo

Giulio Carcano

*** ***

Cesare Guasti (1822-1889), writer and philologist, was secretary to the Accademia della Crusca. Politically he was a supporter of the conservative and Catholic movement.

In his letter he admits that he is not an admirer of Cicero, but complains that Mommsen offended not only Cicero, but also the Italians, defining them as lacking any artistic, poetical and musical quality.

Di Firenze, il 18 giugno 1879

Egregio signore,

Vorrei meritare la stima che Ella fa di me; ma sento pur troppo di non essere qual Ella mi crede. Se per altro da quella stima nascono la benevolenza e l'amore che pur mi dimostra, condono volentieri l'errore, e accetto la Sua cara amicizia. Della quale mi è pegno il volume che mi ha regalato, piccolo di mole ma grande nel pensiero. Cicerone anche dai nostri fu giudicato severamente; né io sono ammiratore dell'uomo, quantunque i suoi scritti legga ammirando. Ma il Tedesco ha offeso qualcosa che non è Cicerone;[24] e il traduttore del Mommsen ha fatto più grave lo sfregio con la barbara frase che non oso chiamare italiana. Tutto si può dire, ma sta nella misura e nel modo. La critica non ha più oggi pudore: ed è più lodata, quanto più alto percuote, e più fa demolire le glorie antiche. Non si aspetti dunque, mio buon Signore, d'esser lodato dall'aver preso a difendere Cicerone: ma per questo io apprezzo tanto più l'animo Suo generoso, mentre lodo nel Suo libro l'ingegno e il sapere.

Sono il Suo
Aff.mo ed Obb.mo
Cesare Guasti

*** ***

Atto Vannucci (1810-1883), patriot and historian, the author of *I Martiri della Libertà Italiana dal 1794 al 1848*, *Storia d'Italia dall'Origine di Roma all'Invasione Longobarda*, *Storia dell'Italia Antica.*

Vannucci was at the time severely ill, and, while he informed others that he was not in a condition to respond, he wants to answer M. Messina personally, congratulating him on his performance.

Firenze, 8 marzo 1879

Pregiatissimo Signore,

I dolori di questi ultimi tempi m'impediscono di far cosa che potesse esser piacevole agli altri e a me. Pure in qualche momento di respiro tentai di leggere il libro da Lei favoritomi, e con molte interruzioni lo vidi, e ne ebbi buona impressione. Solo mi ricordo di questo. Ora m'è la vita continuamente sconvolta. Ad altri ho fatto

---

[24]L'illustre Segretario dell'Accademia della Crusca, commendator Guasti, allude alle oltraggiose parole, che il Mommsen rivolse agl'Italiani in generale, reputandoli sforniti di ogni facoltà, poetica, artistica, e segnatamente musicale: vedi Mommsen, *Stor. Rom.* Traduzione di Giuseppe Sandrini, Torino 1857, Tom. I, lib. I, cap. XV, pag. 199 (note by Michele Messina).

sapere che non ho forze bastanti a rispondere. A Lei mando queste poche parole, e prego la Sua benignità a volermi scusare.

Accolga i miei saluti e mi creda

Dev.mo Suo

Atto Vannucci

*** ***

Ercole Ricotti (1816-1883), professor of Modern History at the Univ. of Torino, President of the Academy of Sciences of that city, Senator of the Kingdom, was the author of the *Storia della Monarchia Piemontese*, in six vols.

Ricotti observes that Mommsen is sometimes drawn to 'strange ravings', such as when he denies that the Italians have 'poetic, artistic, and much more musical faculties, when he alleges that Dante is just a rhetorician, and establishes the Apennines as the boundaries of Italy'.

Torino, 30 marzo 1879

Pregiatissimo signore,

Ho letto colla massima attenzione il libro, di cui Ella mi volle generosamente gratificare. Ne ho trovato ottimo ed opportuno lo scopo, molta la dottrina, chiara l'esposizione. Forse, se ci apparisse minor passione, l'effetto sarebbe più profondo. Ma Ella certamente è giovane, e ne ha coi pregi anche i difetti: dei quali si curerà pur senza volerlo di giorno in giorno. Benedetta gioventù!

Del resto, Ella ha fatto opera buona e degna del grande avversario che si è scelto. Il Mommsen, infatti, se ha immensi pregi, si lascia talora trascinare dall'ingegno e dalla solitudine a vaneggiamenti strani, come quando nega agli Italiani la facoltà poetica, artistica, e più la musicale, e dà del retore a Dante, e stabilisce all'Appennino i confini di Italia. Amor di patria vuole che questi trascorsi d'una gran mente sieno ribattuti a dovere: ma la calma imparziale fa miglior giovamento.

Intanto mi congratulo con Lei e La ringrazio.

Suo devotissimo

Ercole Ricotti

*** ***

Luigi Tosti, patriot and historian, was the abbot of the Abbey of Monte Cassino. A leading figure in the world of learning of the Risorgimento Italy, and an expert in German historiography, he was the author, among other works, of the *Storia della Lega Lombarda.*

Abbot Tosti congratulates M. Messina for defending Cicero against the <contumelies> of Mommsen, and adds that "criticism is healthy in the limits of reason, pestilential if boundless".

Napoli, 18 febbraio 1879

Pregiatissimo Signore,

Ho ricevuto qui in Napoli, dove sono da qualche tempo, il gentile dono che mi ha fatto della Sua Apologia di Cicerone, e la Sua lettera.

Le rendo infinite grazie della cortesia che mi usa, e Le significo le mie sincere congratulazioni. Il Suo libro è stato veramente opportuno, e debbono essere a Lei riconoscenti quanti amano le lettere latine, avendo con molta erudizione e critica difeso il più grande oratore della Repubblica Romana dalle accuse, o meglio, contumelie del Mommsen. La critica è salutare nei confini della ragione; pestilenziale se sconfinata.

Grazie dunque di nuovo, e faccia altri libri, che come questo accresceranno la stima verso di Lei del

Suo dev.mo servitore

Luigi Tosti

*** ***

Augusto Conti (1822-1905), patriot, philosopher, professor in Pisa and Florence, a member of the Accademia della Crusca. His numerous philosophical works are known more for the elegance of their style than for their soundness of thought.

Conti writes that love for his own country made M. Messina 'indignant at so base contumelies towards a man of so much glory among us', and the honesty of his soul led him to defend 'the abused name of that great man of ours'.

Firenze, 26 di giugno 1879

Degnissimo Signore,

L'Apologia di Cicerone, scritta con eloquenza, con erudizione vera, con alto affetto, con efficace dimostrazione del Suo argomento, è un'opera di buon cittadino e di onest'uomo. L'amore della patria rese Lei sdegnoso a contumelie così turpi verso un uomo di tanta gloria fra noi; l'onestà dell'animo La invitò a difendere il nome oltraggiato di quel nostro grand'Uomo. I morti ancora dobbiamo difendere, perché l'anima loro è viva. O Signor mio, quel Dottore là mi sembra un frenetico. Maledetta gelosia! E quando mai saremo nel Regno della verità e della carità?

Suo

Augusto Conti

*** ***

Pietro Fanfani (1815-1879), classical philologist and patriot, the author of the *Osservazioni al Nuovo Vocabolario della Crusca* and of the *Vocabolario della Lingua Italiana.*

Fanfani complains that, unfortunately, people like Mommsen are found not only in Germany, but also in Italy. In particular, in Florence prof. Trezza, in a lecture on Cicero, not only accepted the opinions of Mommsen, but used his mockery. Therefore, true criticism has in Italy just a few worshippers.

Firenze, 16 febbraio 1879

Caro signor Messina,

La Sua Apologia non solo mi è sembrata un lavoro di eletta dottrina, e (non) di peregrina erudizione; ma un lavoro altresì di critica efficace e calzante. Ma disgraziatamente i Mommsen non sono solo in Germania; e piglia piede anche qua l'audacia critica de' Tedeschi: anzi, rispetto a Cicerone, qui a Firenze, all'Istituto di Studi Superiori, il prof. Trezza[25] fece una lezione, dove non pure accettò le opinioni del Mommsen, ma andò molto più in là, giungendo fino allo scherno.

Né il Trezza è solo, in Italia; e i giovani abboccano avidamente, per poi ripetere le medesime cose quando saranno professori essi. Tali eccessi portano poi un altro danno gravissimo; ché i paurosi, spaventati, precipitano nell'eccesso contrario, mantenendo tutti i giudizi falsi dati dai nostri vecchi, e non volendo riconoscere la verità, anche palpabile, anche in certi casi, dove, come in quello di Dino Compagni,[26] la critica scuopre e dimostra le più strane imposture. E così la vera critica ha in Italia pochi cultori. Speriamo nell'avvenire.

Mi resta solo il ringraziarLa caramente, e il ricordarmeLe.

Suo leale servitore

Pietro Fanfani

*** ***

---

[25] Gaetano Trezza, patriot and philologist, professor of Latin Literature at the Istituto di Studi Superiori in Florence.

[26] Dino Compagni (1255-1324), author of the *Cronica delle cose occorrenti ne' tempi suoi*. Fanfani in his *Dino Compagni vendicato dalla calunnia di scrittore della cronaca* and in other writings had tried to demonstrate the falseness of the *Cronica*. The authenticity of the work was demonstrated by Isidoro del Lungo (1841-1927).

Giacomo Zanella (1820-1888), poet, literary critic, prof. of Italian Literature at the Univ. of Padua, translator of Greek and Latin authors, an expert in German and English languages.

Zanella believes that <that German> 'aims at restraining the Latin race, in order to extol his own', exceeding any limit when he defined Dante a 'minor poet'.

Vicenza, 2 luglio 1879

Egregio Signore,

Ella ha fatto quello ch'è vergogna non abbiano prima fatto i nostri letterati di mestiere. Non può credere quanto piacere mi abbia cagionato il Suo libro, ove una scelta erudizione è congiunta a tanto calore di stile. Io credo che quel Tedesco cerchi di reprimere in generale la razza latina per esaltare la sua. Che cosa non ha poi detto di Dante poetucolo? Io La ringrazio a nome della patria comune e mi segno

Di Lei Devotissimo

Giacomo Zanella

*** ***

Emanuele Rocco (as spelt in the *Dizionario Treccani*), classical scholar, translator of Greek and Latin classics, and of Émile Zola into Italian.

Rocco maintains that German scholars have 'a craving for paradox', often use manuscripts of doubtful authority, and presume to raise doubts about the greatness of our glorious men.

Napoli, 18 maggio 1882

Mio egregio amico,

Ho letto con molto piacere la vostra Apologia di Cicerone e i giudizii, che ve ne hanno dato tanti valentuomini. Questi signori Tedeschi, a cui molto dobbiamo dopo che lo studio dei classici greci e latini passò dall'Italia fra loro, hanno un pò la smania del paradosso, e spesso fanno

verificare quel nostro detto che chi si assottiglia si sca-
vezza. Cominciarono col guastare i testi, cavando dalle
loro biblioteche un numero infinito di manoscritti di ben
dubbia autorità, le cui varie lezioni ottenebrano le cose
più chiare. Vennero poi a porre in forse l'autenticità di
molte opere, che dagli antichi sono a noi pervenute, e
quasi rinnovarono le pazze opinioni del p. Arduino.[27]
Ora non contenti di ciò vogliono entrare a discutere il
merito di quelle opere, e naturalmente, per dir cose nuo-
ve, cadono nello strano e nello stravagante. Io non credo
che lo facciano per invidia che abbiano delle nostre glorie; ma è certo che essi ci danno ragione di crederlo. E
però il vostro lavoro non può non riuscir caro a quanti
hanno una religiosa venerazione, per cotesti nostri mag-
giori, da cui traggono la vita quanti pregevoli scrittori
vanta il mondo civile. Furono uomini e quindi s'ebbe-
ro i loro difetti, ma furono uomini prodigiosi quando si
consideri i tempi, in cui vissero e lo stato in cui erano e
le scienze e le lettere allorché vennero al mondo.

Mi congratulo adunque con voi, e tanto più vi trovo
degno di lode, in quanto che avete dovuto combattere
contro uno di coloro, che vanno per la maggiore e che
ha in altro modo reso importanti servigi alla classica
filologia.

Vogliatemi bene e credetemi

Vostro aff.mo amico

Emmanuele Rocco

*** ***

---

[27] Father Arduino, a Jesuit, who maintained that the works of the Fathers of the
Church were only supposed to be so, but, accused of heresy, was obliged to
recant. He was refuted by Maturinus Veissière de la Croze in his *Vindiciae
Veterum Scriptorum contra Johannem Harduinum* (Rotterdam, 1708).

Appendix B

From the *Enciclopedico*

Apologia
di
Cicerone
contro
Teodoro Mommsen
per
Michele Messina
Napoli, Tip. Eugenio 1878

The reviewer writes that the <supreme> philosopher Cicero is held in contempt by Mommsen, by whom he is considered as a 'vulgar man', and so too the Italians, who follow the latest fashion of the minute, foreign rubbish, who praise him. Nevertheless Mommsen judges with a spirit blinded by passion, and when he cannot find a personality in Germany who is able to surpass one of another nation, he tries to detract from his merits. Therefore, his *Roman History* is 'highly inspired by a partisan spirit'.

> Il padre dell'eloquenza latina, il sommo oratore, il politico, lo scrittore, il filosofo sommo Marco Tullio Cicerone è fatto segno di disprezzo, è considerato come uomo volgare, anzi peggio da Teodoro Mommsen; e gl'Italiani amanti delle novità del foresterume l'applaudiscono- Chi è Cicerone? lo sanno tutti: è il maestro dei Maestri dell'eloquenza, l'avvocato per eccellenza, è il giureconsulto mondiale.

> Teodoro Mommsen pure è conosciuto; egli addottorato in vari rami dello scibile umano, erudito archeologo scrittore forbito, ma non giusto, egli giudica con lo spirito di passione, e quando nella sua Germania non rinviene un personaggio, che nel parallelo superi quello di un'altra nazione, cerca di menomarne i meriti, di avvilirlo.

> Per quanto rispetto godiamo del Mommsen e di tutta la Germania, non possiamo astenerci dal rivendicare i propri diritti; dichiarando quindi eminentemente partigiana

la Storia Romana del Mommsen, elogiamo il signor Michele Messina che con sodi argomenti, con autorità accertate dichiara Cicerone quel Cicerone che fu e che da tutti generalmente è ritenuto, confutando esattamente le parole del Mommsen.

Il Messina merita la riconoscenza degl'Italiani, perché ne ha saputo rivendicare l'onore, e noi in nome di tutti gli tributiamo le più sentite azioni di grazie, ed i ben meritati encomi.

(Dall'*Enciclopedico*, periodico di scienze, lettere, arti e filantropia, fondato e diretto dal Comm. Prof. Giuseppe Barbieri da Larino.

Larino- Aprile a Luglio 1879- Num.11-12, pag. 1058-1059)

*** ***

From the journal *Propugnatore*

MESSINA MICHELE
Apologia di Cicerone contro Teodoro Mommsen
Napoli, Stab. Tip. di A. Eugenio 1878, in 8, di pag. 173

The reviewer criticizes the 'insolent arrogance' of the critics on the other side of the Alps, and complains that, in his own times, the Italians admire and flatter anything that comes from abroad.

Il signor Michele Messina, che ebbe la santa idea di difendere il Principe della Romana Eloquenza contro il suo denigratore Teodoro Mommsen, che fece della sua opera una sdegnosa e nobile protesta contro l'insolente burbanza dei critici d'oltr'alpe, divise in cinque ragionati capitoli il suo lavoro, premettendovi un assennato discorso. Considerò nel primo capitolo Cicerone come politico, esponendo i pregi del libro della Repubblica, fonte inesausta di morale e politica dottrina, difendendolo dall'accusa di versatilità lanciatagli mal a proposito dal Mommsen, tessendo la giusta storia della sua

condotta come cittadino e come uomo di Stato. Nel
secondo capitolo presentò Cicerone scrittore, sostenen-
done la grandezza dell'ingegno, illustrandone le opere e
riportandone i giudizii che molti grandi uomini hanno
dato di lui grandissimo. Nel terzo capitolo difese l'oratore, esponendo i pregi della sua eloquenza. Nel quar-
to pose Cicerone a confronto de' suoi contemporanei,
esaltandone con molta evidenza l'immensa superiorità
non solo, ma riportando anche, a proposito dei suoi de-
trattori d'allora, il giudizio degli scrittori dell'epoca, e
giustamente da ultimo osservando che colla morte di Cicerone decadde la vera Romana Eloquenza. Nel quinto
ed ultimo capitolo l'Autore considerando Cicerone come
filosofo, ne dichiarò gli intenti, ne illustrò l'erudizione,
ne esaltò giustamente la filosofia diretta alla conoscenza
di una Divinità unica, eterna, universale, creatrice del-
l'anima immortale, giusta distributrice dei premi e delle
pene, e con una nobile perorazione coronò degnamen-
te il suo bel lavoro, nel quale seguitando e confutando
Mommsen a passo a passo non sai se più mostri sfoggio
di varia erudizione, o possanza di irrefutabili argomenti.

Bel lavoro! E lo ripeto col più profondo convincimento.
Ma anche e sopratutto, buona azione. Ai nostri tempi
in cui si ammira tutto ciò che viene dall'estero, anche
quando attenti alle più incontrastate, alle più luminose
glorie del nostro paese, ai nostri tempi in cui da una
turba di adulatori si incensa ogni gloria d'oltremonti,
purché camuffata di una toga, e preceduta da un gri-
do di celebrità, per quanto discutibile sia; è raro, ma è
pur confortante il trovare, un uomo dotto ed onesto che,
per rispetto delle antiche glorie della nostra Genitrice,
ardisca di sorgere ad impugnare le dottrine con tanta
asseveranza bandite. E quest'uomo dotto ed onesto è il
signor Michele Messina; e la sua difesa è legittima, la sua
ira è giusta (lo dichiara egli e io lo confermo), la tenzone
è pietosa e, non che degna di scusa, da commendarsi ol-
tremodo, in omaggio alla memoria del grande Arpinate,
che da duemila anni a ragione onorano tutti gli uomi-

ni, infiammati dall'amor della scienza e dal desìo della gloria.

A. B.

(Dal *Propugnatore*. Studii filologici, storici e bibliografici in appendice alla collezione di opere inedite o rare di vari socî della commissione pei testi di lingua.

Anno XII. -Dispensa 1 e 2- Gennaio, Febbraio, Marzo, aprile 1879- Bologna).

*** ***

Giovanni La Cecilia (1801-1880), patriot and writer, known in France as Jean La Cécilia, suffered 26 years in prison and exile. He published his political writings both in French and in Italian.

La Cecilia writes that Mommsen is the disfigurer of Cicero's wisdom, and condemns his effrontery in judging him. He uses harsh words, when he speaks of the 'insane judgment of a descendant of the traitor Arminius and of the marauders of Barbarossa', curiously adding: 'Twenty centuries ago Cicero saved Rome from the rage of Catiline .... find one who resembles him, if you want to save your Berlin from the fury of your fellow-citizen socialists and anarchists'.

Apologia di Cicerone
contro
Teodoro Mommsen
per
Michele Messina
Napoli- Tipografia Eugenio, 1878

Difendere Cicerone! Ma sembra un paradosso, dopo secoli di non mai ininterrotta tradizione, che lo paragonava soltanto al fulgido astro della Grecia: Demostene!!

No, non è possibile, un detrattore del nostro grande Arpinate sarebbe stato, se non lapidato, almeno scacciato tra sibili acuti dalle terre italiane, come profanatore delle nostre glorie.

Morirò dunque fra le illusioni, ripetei a me stesso, appena sfogliato il libro dell'Apologia di Cicerone, pubblicato dallo egregio giovane Michele Messina, che alla profonda dottrina accoppia l'ingegno e l'amore indomito verso la vera grandezza della patria nostra.

Il deturpatore della sapienza e della eloquenza popolare e forense di Cicerone non solo vive ed esiste in Italia, pregiato ed onorato da tutti coloro, che oggi preferiscono le nebbie germaniche al sole dell'antico Lazio ed alla italica filosofia; ma è professore dell'Università di Berlino, è Mommsen che vantasi dottissimo dei fatti di Roma, per iscriverne la storia ed aver l'impudenza di emettere un giudizio sull'illustre Arpinate, che suona così:

Cicerone politico.

"Ci fu mestieri di parlare già parecchie volte di quest'uomo eruditissimo. Come uomo di stato, senza penetrazione, senza opinioni e senza mire (sic): esso ha successivamente figurato come democratico, come aristocratico e come strumento dei monarchi, e non fu giammai altro, che un egoista di vista corta".[28]

Che dire su quest'insano giudizio d'un discendente del traditore Arminio, o dei predoni del Barbarossa, squadernarli dinanzi, come ha fatto l'egregio Messina, tutte le storie, tutti gli autori della vecchia e nuova età, che pregiarono e giudicarono il grande oratore di Roma come sagace politico ed incomparabile filosofo, a cominciare da Sant'Agostino e Tertulliano e terminare al greco Plutarco e ad Erasmo di Rotterdam.

E perché nulla mancasse all'Apologia di Cicerone, il Messina non ha lasciato di presentarcelo, anche Dei-

---

[28]Mommsen, *Storia Romana.* Traduzione di Giuseppe Sandrini (Milano, Guigoni, 1865), tom. III, lib. V, cap. XII, pag. 582 (note by M. Messina).

sta col suo libro *De Natura Deorum*, esclamando: "Egli sciolse la grave questione dei veri beni e dei veri mali, e sullo scopo ultimo di tutte le azioni umane, egli finalmente fu il solo in quell'epoca, che, persuaso della falsità d'una religione ridicola professata dai Romani, si elevò coi lumi della filosofia alla conoscenza di una divinità unica, eterna, universale, provvida, ed a riconoscere l'immortalità dell'anima, ed una vita futura di premio e di pena."

Michele Messina, nel pubblicare l'Apologia di Cicerone, non solo faceste conoscere il vostro ingegno e la vostra erudizione, ma compiste una buona azione.

Signor Mommsen, venti secoli or sono, quel politico, che avete calunniato, salvava Roma dai furori di Catilina e proteggeva i destini dell'umanità; cercatene un altro che a lui somigli, se volete salvare la vostra Berlino dal furore dei vostri concittadini socialisti ed anarchici; sarà difficile, perché i Ciceroni nascono soltanto su questa terra del Genio e della Luce".

Giovanni La Cecilia

(Dal giornale *La Discussione*, n. 304, 3 novembre 1879)

*** ***

On p. 152 of the book by M. Messina is printed the advertisement, that I transcribe here:

<Soeben erschien: Apologia di Cicerone contro Teodoro Mommsen per Michele Messina 8. (173 Pag.) Preis 3 fr.

Dieses scharfe Libell gegen den berühmten Gelerten wird Aufsehen erregen und hoffentlich widerlegt werden>.

AKADEM. BUCHHANDLUNG

(Dal Börsenblatt di Lipsia, 27 dicembre 1878, n. 299).

Si è pubblicato: Apologia di Cicerone ecc. Questo acuto libro contro l'illustre scienziato farà romore: speriamo che sia confutato. Libreria Accademica.

*** ***

Paul Guillaume (1842-1914), historian and abbot, spent many years in Italy, writing the *Description historique de Mont-Cassin* and the *Essai Historique sur l'Abbaye de Cava*. He returned to France in 1878.

In his letter he praises Michele Messina for his performance, and writes that Mommsen is 'a jealous and cold historian'.

Champigny (Seine) 27 janvier 1879

Très-honoré Monsieur Messina,

L'Apologia di Cicerone, que vous avez bien voulu m'adresser en France, est venue me trouver aux environs de Paris, où je réside actuellement.

J'ai lu d'un trait votre travail. Il m'a grandement intéressé; et je tiens, tout d'abord, à vous offrir mes sincères félicitations.

Je connaissais les nombreuses attaques de Mommsen contre votre illustre orateur d'Arpinum, attaques qui ont déjà fait un certain bien dans le monde des lettres; mais je ne connaissais point encore de réplique, un peu sérieuse, à tant de propositions si effrontément avancées.

Après la lecture de votre Apologia, fruit de longues méditations et de beaucoup de recherches sérieuses, le doute sur les qualités et les mérites de Cicéron n'est plus possible. Qu'on le considère, avec vous, Monsieur, comme homme politique ou comme écrivain, comme orateur ou comme philosophe, en face de ses contemporains ou devant la postérité, toujours ou partout on trouve grand

quel Marco Tullio.....

....

<Di cui la fama ancor nel mondo dura

E durerà, quanto 'l mondo lontana>.[29]

Peut-être y en aura-t-il qui critiqueront dans votre livre un trop grand luxe de citations latines ou grecques, italiennes ou françaises. Á mes yeux, c'est un mérite de plus. Il est beau de voir tant d'écrivains rendre hommage à la vérité, et venir appuyer de leur autorité les arguments, à la fois si solides et si finement spirituels, que vous opposez aux allégations du jaloux et froid historien allemand. J'applaudis, Monsieur, au sentiment patriotique qui vous a inspiré la belle Apologia di Cicerone contro Teodoro Mommsen, et je fais des vœux ardents pour que tous les lecteurs de l'historien allemand puissent lire aussi la spirituelle et savante réplique du littérateur italien.

Je suis avec des sentiments d'admiration, Très-honoré Monsieur,

votre très-obligé et bien respectueux serviteur

Paul Guillaume

*** ***

François Tommy Perrens (1822-1901) was the author of *Jerôme Savonarole, Histoire de Florence*, in six vols, *La Civilisation Florentine du XIIIe au XVIe siècle, Histoire de la Littérature Italienne.*

In his letter he assures his correspondent that he shares his opinions about Mommsen, whom he accuses of having a 'disorderly mind', that he committed 'many errors and insanities', and he also blames him for the 'contemptible portrait he gave of the Italians'. He adds that Mommsen enjoyed from Napoleon III a pension of 10,000 francs and, having lost it after the Franco-Prussian war of

---

[29] <whose fame /yet lives, and shall live long as nature lasts!>, Dante, *Inferno*, II, 59-60 ; translation by Henry Francis Cary. The two lines of Dante refer here to Virgil, and are cited by M. Messina on p. XIII of his book.

1870, 'he did not refrain from writing about France, then overcome and in chains, the most shameful words that could ever come out from the pen of a man of letters'.

Perrens declares himself 'an implacable enemy of the Teutonic stock'. Although he recognizes 'the primacy of the Germans in the sphere of learning, and probably of philosophy', he cannot admit 'that we have to take from them lessons in the art of composing a book ... In Italy you have two literary schools: the one that follows the frivolous French of the past, and performs worse than them; the other that follows the grave Germans of the present, and performs better than them'. Therefore, he disagrees with the admiration of Messina's preface for the literary pre-eminence of the 'German rubbish', and he considers it just a 'rhetorical precaution', that allows him better to tell the truth.

ACADÉMIE de Paris, Université de France

Paris, le 4 juillet 1879.

Rue Scheffer, 7.

Pregiatissimo Signor Professore,

Fu per me una ben certo inaspettata ventura l'aver ricevuto il preziosissimo regalo del Suo libro, vuoi per la materia ivi trattata, vuoi per l'onore che mi vien fatto, e per quello di esser conosciuto e stimato costà dai dotti quale Ella è. Bench'io forse lo meriti pel vivissimo amor d'Italia e pel desiderio di servirla, pur troppo so che non basta amare e servire; la Sua lettera, egregio signor Professore, ed il Suo omaggio mi lasciano credere che non ho servito troppo male, poiché le mie povere cose non sono del tutto sconosciute nella vostra splendidissima penisola. E poi si direbbe daddovero che Ella abbia indovinato quali siano i miei sentimenti, le mie opinioni, ché di fatto io partecipo assolutamente le Sue attorno al Cicerone e anche al Mommsen; e non potevo altro che rallegrarmi di vederle espresse in tanto bello stile, in tanta armoniosa lingua. Né contro il solo oratore romano si spacciò in ismisurati capricci lo storico tedesco, ma quanti errori, quante insanità commise la sua sregolata mente, sarebbe lungo a dirlo. Io volentieri limiterei le mie lodi al suo primo volume nel quale riduce quasi al vero la leggenda dei Re e dei primi tempi della Repubblica; ché per il rimanente pare un tutt'altro uomo. Né mi par possibile il perdonargli l'indegno ritratto cui egli dipinse degl'Italiani, e io spero che anche sopra siffatto punto potremo andar concordi. Come, quanto a me, dimenticherei che il medesimo, godendo dall'Imperatore Napoleone III una pensione di ben diecimila franchi, avendola perduta colla guerra del 1870, ebbe l'ardire di sollecitarne la continuazione dall'Accademia delle Iscrizioni, e questa avendogli risposto che le mancavano denari, per sostituirsi ai Mecenati scacciati, egli non dubitò di scrivere intorno alla vinta e incatenata Francia le più turpi parole, che siano mai uscite dalla penna di un letterato![30] Sono questi due fatti innega-

---

[30] On the same subject see here below, text corresp. with n. 31, letter by Paul Albert to M. Messina.

To this allegation, recurring in the French press, Mommsen answered with harsh words: see his 'In eigener Sache', *Reden und Aufsätze* (Berlin, 1912), pp. 427-431.

He admitted that someone had suggested a collaboration with Napoleon's *History of Julius Caesar*, but he refused to be involved, and never wrote a pen-stroke ("ich habe für die Schriftstellerei des Kaisers nie einen Federzug getan"). Notwithstanding such a refusal, Napoleon gave him permission to consult any manuscripts in the library of Paris without further official request, and this permission was particularly useful to him for his works ("trotz jener tatsächlichen Ablehnung ... jede Handschrift der Pariser Bibliothek ohne offizielle Vermittlung direkt erbitten zu dürfen, was mir bei meinen Arbeiten von wesentlichem Nutzen gewesen ist"). He had been accused of getting a pension from Napoleon, but he had never received any money from the French treasury nor from the private casket of the Emperor ("ich habe niemals weder aus einer französischen Statskasse noch aus der kaiserlichen Privatschatulle Geld empfangen").

During the Franco-Prussian war he wrote his *Letters to the Italians*, urging them not to take part with France, but he did it only after long hesitation, certainly not light-heartedly, and because it was his duty as a Prussian subject. Nevertheless he complained of the enmity between the two nations. For Mommsen and the French world of learning see J. von Ungern-Sternberg, 'Theodor Mommsen und Frankreich', in *Francia*, 31, 3, 2004, pp. 1-28; 'Deutsche und Französische Altertumswissenschaftler vor und während des ersten Weltkriegs', in H. Bruhns-J.M. David-W. Nippel eds., *Die späte Römische Republik. La fin de la République Romaine*, Collection de l'École

bili, ancorché li rammenti qui un (implacabile nemico della stirpe teutonica), per usare le parole da Lei dirette (ai cortesi lettori). Dovrebbe forse tale sentimento interdirmi la licenza di oppormi al Suo sopra la preminenza letteraria del tedescume in questo secolo. Mi permetta però di dichiararLe, che riconoscendo il primato dei Tedeschi nella sfera dell'erudizione, e forse della filosofia, ancorché per l'erudizione abbiano emuli in Italia e in Francia, e per la filosofia in Inghilterra, non posso ammettere che noi abbiamo a prendere da loro lezioni nell'arte di comporre un libro, di narrare, di scrivere; il che non è una parte di mediocre polso nella letteratura. Voi in Italia avete due scuole letterarie: l'una che segue i frivoli francesi del passato e fa peggio di loro; l'altra che segue i gravi Tedeschi del presente e fa meglio di loro. Io capisco dunque benissimo l'ammirazione che spunta nel Suo proemio, la quale è forse, più che altro, una precauzione oratoria nell'atto di dire le Sue verità (temo che la parola non sia francese più che italiana) ad uno dei più rinomati storici di colà.

Ma che fo io così favellando, senza badare che vo disponendo del tempo di V.S.! E questa sia almeno una prova del piacere che mi fece, favorendomi il Suo generoso libro e procurandomi l'occasione di dichiararmi qual sono

Di Lei, pregiatissimo signor Professore

Tenutissimo e devotissimo servo

François-Tommy Perrens

*** ***

Française de Rome n. 235 (Rome, 1997), pp. 45-78; 'Mommsen in Frankreich: Übersetzungen und Rezensionen', in Ch. Avlami-J. Alvar, eds., *Historiographie de l'Antiquité et Transferts Culturels*, Rodopi (Amsterdam-New York 2010), pp. 285-299.

In a private communication to the present author J. v. Ungern-Sternberg insists that the allegations against Mommsen are "vollkommen unbelegt", quite unproven.

Paul Albert (1827?-1880), professor of Langue et Littérature Française Moderne at the Collège de France, author of a *Histoire de la Littérature Romaine* in two vols (Paris, 1871) and of several books on Modern French Literature.

He writes that Mommsen does not care for the truth, but "collected from any side arguments that could confirm his own opinion. With the same intention he had begun to narrate Roman events, in order to demonstrate that Caesar had not usurped power, but that he exerted it necessarily and legitimately. For that reason he dealt with any possible affront Cato, Cicero, Brutus, finally, all the defenders of the laws and of liberty". Furthermore his work "pleased Napoleon III so much, that he associated him in writing his own history of Julius Caesar, and enriched him with a lot of money. Disappointed for having lost it, he now inveighs with harsh words against our republic".

Paris, 8 juillet 1879

Grato animo, doctissime Professor, librum tuum accepi. Ipse ego abhinc ante decem annos, historiam litterarum romanarum scribens, quid sentirem de iniquo Mommsenii judicio explicui; sed tuum erat et quasi gentilium munus causam Ciceronis suscipere et singulari affectu perorare.

Mihi videtur Mommsenius not ita quid verum esse curavisse quam argumenta ex omni parte collegisse quibus prejudicatam opinionem confirmaret. Eo enim consilio res romanas narrandas susceperat ut ostenderetur Caesar non imperium usurpavisse, sed necessarium et legitimun exercuisse. Idcirco omni contumelia affecit Catonem, Ciceronem, Brutum, omnes denique defensores legum ac libertatis. Sic opus ejus placuit Napoleoni III, qui eum sibi socium assumpsit in scribenda Caesaris historia, et magna pecunia locupletavit, quam nunc amissam dolens, in rempublicam nostram acerrime invehitur.[31]

---

[31] For the same subject see here above, letter by François-Tommy Perrens, text corresp. with note 30.

Liceat mihi Franco Gallo, ut dicunt Germani, reipublicae semper amantissimo, philosopho, eam operis tui par-
tem minus laudare, qua religioni tuae gratificans, cla-
rissimos viros, libere sentiendi duces, Voltaire, Diderot,
acerbo, ut mihi videtur, judicio premis. Hos admiraban-
tur, hos imitabantur quod poterant Germani isti quos
tantis laudibus extollis. Gloriosum sibi duxit Wieland
(non Lessing, ut putas) cognomen Voltairii Germani,
quod quidem qua ratione meritus sit parum video.

Hactenus de istis. Malo tecum de Cicerone consentire,
licet non iisdem de causis, quam in quibus dissentiam
referre.

Honoratissime collega, Salve. Vel potius χαῖρε.

Paul Albert

Professeur au Collège de France

Quai de Béthune, 24

*[Very learned professor, with grateful feelings I have re-
ceived your book. Ten years ago, while writing my* History of Roman Literature, *I explained my opinions about
Mommsen's unfair judgment; but it was up to you almost as a duty on the part of a fellow countryman to
defend the good reasons of Cicero, defending them with
great affection.*

*Mommsen appears to me as not caring for what is the
truth, but as having collected from any side arguments
that could confirm his own opinion. With the same intention he had begun to narrate Roman events, in order
to demonstrate that Caesar had not usurped power, but
that he exerted it necessarily and legitimately. For that
reason he dealt with any possible affront Cato, Cicero,
Brutus, finally all the defenders of the laws and of the
liberty. His work pleased Napoleon III so much, that
he associated Mommsen in writing his own history of
Caesar, and enriched him with a lot of money. Now,*

suffering grief for having lost it, he inveighs against our republic with harsh words.

But let me a Franco-Gallo, as the Germans call us, always devoted to the Republic, and a philosopher, praise less that part of your work, in which, showing kindness to your own religion, you judge bitterly celebrated men, like Voltaire, Diderot, the masters of free thought. These Germans, that you so highly praise, used to admire and to imitate them as much as they could. Wieland -not Lessing, as you believe- took the surname of the German Voltaire, although I scarcely see the reason why he merited it.

However, enough with them. I prefer to agree with you about Cicero, rather than to refer to subjects on which I disagree.

Very honourable colleague, greetings! or, rather, χαῖρε.[32]]

*** ***

Felix Bouchot, prof. de rhétorique, author of a *Précis de Littérature Ancienne*, editor of Virgil.

He writes that 'only a German' could violate the memory of Cicero, and adds that the 'guiding principle' of the Germans is, that "they criticize what they admire, and they admire what others criticize: and in this way they claim to say new things". Bouchot nevertheless disagrees on the point that Cicero was an outstanding politician.

Parisiis, mense octobre MDCCCLXXIX

Illustrissime Professor,

Cum jam me tardiorem in meritis tibi gratiis agendis, quod tuam Ciceronis apologiam mihi miseris, nolim officii mei omnino incuriosum haberi.

Jamdiu tibi scribere debuissem atque manifestum facere quanta cum voluptate legerim hunc libellum, tum

---

[32]Translation by author.

sententiarum gravitate, tum verborum elegantia expolitum. Non tantum urbane, sed etiam recte apteque mihi
disputasse de ea re videris. Quis velit nisi Alemannus,
summi illius viri, Ciceronis memoriam violare? Quis ne-
gare audeat illum fuisse absolutissimum oratorem qui in
philosophia, in concionibus orationibusque, in peroran-
dis causis, in describendis Rhetoricae legibus, excelluerit
ita ut nullum ei opponere scriptorem liceat, nedum Asi-
nium Pollionem[33] e tenebris suis excitare fas sit ut Tullio
comparetur? Quod si attenderis, videris eam esse Ale-
mannorum rationem ut quae mirantur vituperent, quae
vituperant alii mirentur: quo pacto prae se speciem fe-
runt nova dicendi. Quamvis tamen in omnibus coeteris
rebus tecum sentiam, non libenter concesserim Ciceronem summum, quem nunc politicum virum vocare sole-
mus, fuisse. Tantummodo confitebor eum in illis rerum
angustiis fuisse versatum ut facilius sit dicere quid non
agere, quam quid agere debuisset. Vale.

Bouchot

[Paris, October 1879

*Very illustrious professor, although late in thanking you
for your kindness in sending me your defence of Cicero,
I don't want to be believed as paying no attention to my
duties.*

*I should have written to you long ago, telling you with
how much pleasure I read your book, which is at some
points adorned with the dignity of opinions, at some
other points with the elegance of words. You appear to
me as having debated the problem not only politely, but
also accurately and suitably. Who, apart from a German, could desire to violate the memory of that outstanding man, Cicero? Who would dare to deny that he
was the most perfect orator, one who in philosophy, in
meetings and orations, in arguing cases, in describing*

---

[33] Asinius Pollio (75 BC-AD 4), a patron of Virgil and a friend of Horace. As
an orator he was second only to Cicero.

the laws of rhetoric, was pre-eminent to such a degree that it would be worthless to compare any writer with him, still less should one be allowed to rouse from sleep Asinius Pollio in order to compare him with Tully?

If you reflect well, you will see that this is the guiding principle of the Germans, that they criticize what they admire, and they admire what others criticize: and in this way they claim to say new things.

Although, by the way, I am of your same opinion as for the rest, unwillingly I would accept that Cicero was greatest, in what we call a politician. I will only admit that he was inclined to these circumstances of things, in which it is easier to say than to do, at a certain extent, what one should have done. Farewell.][34]

*** ***

Victor Duruy (1811-1894), historian, member of the Académie Française, Minister of Education under the Second Empire, contributed to Napoleon the Third's *Histoire de Jules César*, and was the author of a *Histoire des Romains* in 7 vols.

Duruy writes that Mommsen, 'as many other Germans', has the hand 'harsh and heavy', therefore he was not in a condition of understanding Cicero's 'enchanting spirit'. And Cicero, although sometimes unsteady in politics, remains 'one of the educators of mankind, for the quantity of sound and fortifying thoughts that he, thanks to the magic of his style, has put in the universal circulation'. In defending Cicero, Mr Messina defends 'one of the glories of the Latin stock', whose sons should oppose the 'nebulosity and boldness of the German stock'.

Villeneuve-St -Georges 5 août 1879 (Seine-et-Oise)

Monsieur,

J'ai reçu votre lettre et votre livre. Je vous remercie de l'une et je vous félicite de l'autre. Mommsen, comme

---

[34]Translation by author.

tant d'autres allemands, a la main rude et pesante. Malgré sa connaissance approfondie de la langue latine, il ne pouvait comprendre ce charmant esprit de Cicéron, sa parole harmonieuse don't l'Italie a gardé le secret, et il est dur pour l'un des hommes qui font le plus d'honneur à l'antiquité romaine. Si le caractère n'est pas chez Cicéron à la hauteur du talent, si en politique il a souvent varié, cela intéressait ses contemporains ; pour nous, il est resté un des éducateurs de l'humanité, par la quantité de pensées saines et fortifiantes qu'il a jetées, grâce à la magie de son style, dans la circulation universelle. En le défendant, vous défendez, Monsieur, une des gloires de la race latine. Puissent tous ses enfants rester unis et sauver des nébulosités et des audaces germaniques, le clair génie qu'elle nous a légué.

Je vous renouvelle, Monsieur, tous mes remerci(e)ments.

Votre dévoué serviteur

Victor Duruy

Appendix C: Cicero's birthplace. An open question

According to tradition, Cicero's house was in the place where now stands the Cistercian Abbey of San Domenico, via Ponte dell'Olmo, at that time in the *ager* (country) Arpinas, now part of the administrative division of the city of Sora.

Nevertheless, there is no archaeological evidence that the house actually was in San Domenico, and no conclusion can be drawn, from what we at present know. Therefore let us hear, first of all, from Cicero himself, in his treatise *De Legibus*.

> A(tticus): "Surely I recognize that grove yonder and this oak tree of Arpinum as those of which I have read so often in the '*Marius*';[1] if that famous oak still lives, this is certainly the same; and in fact it is a very old tree".

> Q(uintus): "That oak lives indeed, my dear Atticus, and will live forever; for it was planted by the imagination. No tree nourished by a farmer's care can be so long-lived as one planted by a poet's verses".[2]

> M(arcus):"We agree; and indeed, if you approve, we might walk here by the Liris, in the shade along its bank";[3]

> Plato "discussed the institutions of States and the ideal laws ... in Crete on a summer day amid the cypress groves and forest paths of Cnossus ... we, in like manner, strolling or taking our ease among these stately poplars on the green shady river bank, shall discuss the same subject ...";[4]

---

[1] The poem Marius, written by Cicero in about 59 BC.
[2] *De Legibus*, I, i, 1.
[3] Ibid., I, iv, 14.
[4] Ibid., 15.

A.: "As we have now had a sufficiently long walk, and you are about to begin a new part of the discussion, shall we not leave this place and go to the island in the Fibrenus (for I believe that is the name of the other river), and sit there while we finish the conversation?"

M.: "By all means; for that island is a favourite haunt of mine for meditation, writing and reading".

A.: "Indeed I cannot get enough of this place, especially as I have come at this season of the year.... Hence I used to be surprised (for I had the idea that there was nothing in this vicinity except rocks and mountains, and both your speeches and poems encouraged me in that opinion). I was surprised, I say, that you enjoyed this place so much; now, on the other hand, I wonder that you ever prefer to go elsewhere, when you leave Rome";

M.: "Indeed, whenever it is possible for me to be out of town for several days, especially at this time of the year, I do come to this lovely and healthful spot; it is rarely possible, however ...

To tell you the truth, this is really my own fatherland, and that of my brother, for we are descended from a very ancient family of this district; here are our ancestral sacred rites and the origin of our race, here are many memorials of our forefathers. What more need I say? Yonder you see our homestead as it is now -rebuilt and extended by my father's care; for, as he was an invalid, he spent most of his life in study here. Nay, it was on this very spot, I would have you know, that I was born, while my grandfather was alive and when the homestead, according to the old custom, was small, like that of Curius on the Sabine country. For this reason a lingering attachment for the place abides in my mind and heart, and causes me perhaps to feel a greater pleasure in it; and indeed, as you remember, that exceedingly wise man[5] is said to have refused immortality that he might see Ithaca once more";

---

[5]Odysseus.

A.: "I think you certainly have good reason for preferring to come here and for loving this place. Even I myself, to tell you the truth, have now become more attached to the homestead yonder and to this whole countryside from the fact that it is the place of your origin and birth; for we are affected in some mysterious way by places about which cluster memories of those whom we love and admire ... Therefore in the future I shall be even more fond of this spot because you were born here";

M.: "I am glad, then, that I have shown you what I may call my cradle";

A.: "And I am very glad to have become acquainted with it. But what did you really mean by the statement you made a while ago, that this place, by which I understand you refer to Arpinum, is your own fatherland? Have you then two fatherlands? ... "

M.: "Surely I think that he and all the natives of Italian towns have two fatherlands, one by nature and the other by citizenship ... Thus I shall never deny that my fatherland is here, though my other fatherland is greater and includes this one within it ... But here we are on the island; surely nothing could be more lovely. It cuts the Fibrenus like the beak of a ship, and the stream, divided into two equal parts, bathes these banks, flows swiftly past, and then comes quickly together again, leaving only enough space for a wrestling ground of moderate size. Then after accomplishing this, as if its only duty and function were to provide us with a seat for our discussion, it immediately plunges into the Liris and, as if it had entered a patrician family, loses its less famous name, and makes the water of the Liris much colder. For, though I have visited many, I have never come upon a river which was colder than this one; so that I could hardly bear to try its temperature with my foot".[6]

---

[6] *De Legibus*, II, i, 1, II, iii, 6-7.

Let us hear, now, how many times Cicero in his works and letters mentions Arpinum, and how many times he mentions Sora.[7]

a) Arpinum

"... my farm at Arpinum, which has come down to me from my father and grandfathers ...";[8]

"I mean, he says, that you came from an Italian borough. I admit it and I also add that I am from a town which has twice now brought salvation to this city and her empire";[9]

" ... whenever you come across a man of Arpinum, you will have to listen, willy-nilly, to some fragment of gossip, possibly even about me, but certainly about Gaius Marius ... all that I say about Plancius I say from personal experience; for we at Arpinum are neighbours of the people of Atina ... there was no one at Arpinum, at Sora, at Casinum, at Aquinum, but was Plancius' adherent ... all our rugged countryside, (*aspera et montuosa*) which holds among its hills hearts loyal and unaffected and staunchly true to the bond of kinship ...";[10]

"Marius, born in my native place";[11]

"... the monument of Gaius Marius, saviour of our Empire, afforded his fellow-townsman and defender of the State";[12]

---

[7] I am following the index of the edition Loeb of Cicero's *Works*.

[8] *De Lege Agraria*, III, ii, 9.

[9] *Pro Sulla*, 22-23. "Hoc dico", inquit, "te esse ex municipio". Fateor et addo etiam: ex eo municipio unde iterum iam salus huic urbi imperioque missa est".

The word *municipium* means an Italian town which had full citizen rights. Arpinum acquired these rights in 188. The salvations to which Cicero refers, are the victory of Marius on the Cimbrics and Teutons, and the salvation of the Republic by himself, on Catilina's conspiracy.

[10] *Pro Plancio*, VIII-IX, 20-22

[11] *Pro Sestio*, xxii, 50.

[12] *Pro Sestio*, liv, 116.

" ... Lucius Antonius, whose cruelty I escaped only under the protection of the walls and gates of my native town";[13]

" ... one of the most famous cities of Greece, once indeed a great school of learning as well, would have been ignorant of the tomb of its one most ingenious citizen, had not a man of Arpinum pointed it out";[14]

"... the cool freshness of the streams in my Arpinum";[15]

" ... in this principle (i.e., nature and occupancy) the lands of Arpinum are said to belong to the Arpinates, the Tusculan lands to the Tusculans ...";[16]

"... we shall leave Formiae on the Kalends of May so as to be at Antium on the 3rd ... From there I propose to go to Tusculum, then to Arpinum, returning to Rome by the Kalends of June ...";[17]

"well, I shall go to 'my native hills, the cradle of my birth' ... better the society of countryfolk than of these hypersophisticates";[18]

"we want to stay at Formiae until 6 May. If you don't come before then, perhaps I shall see you in Rome. No use inviting you to Arpinum:

rough land , but breeds good men; and as for me,

I can see no sight sweeter than my home";[19]

"I would go to Arpinum right away upon my word, only it's clearly most convenient for me to expect you at Formiae - up to 6 May ...";[20]

---

[13] *Phil.*, 12.20.

[14] *Tusc.*, V, xxiii, 66.

[15] *Tusc.*, V, xxvi, 74.

[16] *De Officiis*, I, vii, 21;

[17] *Ad Atticum*, 28, (II. 8), Antium, 16(?) April 59.

[18] *Ad Atticum*, 35 (II, 16), Formiae, ca. 28 April 59.

[19] *Ad Atticum*, 32 (II. 11), Formiae, ca. 23 April. The lines, here cited in the Greek original, are from Odyssey, 9.27, with reference to the island of Ithaca.

[20] *Ad Atticum*, 34 (II. 14), Formiae, ca. 26 April 59.

"I mean news of his disgraceful flight and the victor's return - his route and destination. When I hear that, if he is travelling by the Appian way, I am for Arpinum";[21]

"I am staying on at Formiae ... Then I want to go to Arpinum ...";[22]

"The problems you discuss are indeed very difficult ... departure to Arpinum, which might look like flight from Caesar ...";[23]

"... I shall follow your advice not to hide myself away in Arpinum at present, though I wanted to give Marcus his white gown in Arpinum";[24]

"Your admonition not to be too easygoing in what I say (to Caesar) when I see him, to maintain my dignity rather, is excellent ... I mean to go to Arpinum after I have met him, so that I shall not happen to be away when he arrives or be rushing to and fro on a vile road";[25]

"when I have seen him I go on to Arpinum ... what bothers me at the moment is the immediate prospect of meeting him ...";[26]

" ... I will decide whether it is best for me to go to Arpinum or somewhere else. I want to give my boy his white gown, there, I think";[27]

"But I nearly forgot to mention Caesar's disagreeable Parthian shot: If, he said, he could not avail himself of my counsels, he would take those he could get and stop at nothing ... you ask for the rest of the story. Why, he left straight away for Alba, and I am leaving for Arpinum";[28]

---

[21] *Ad Atticum*, 166 (VIII. 16), Formiae, 4 March 49.

[22] *Ad Atticum*, 167 (IX. I), Formiae, 6 March 49.

[23] *Ad Atticum*, 171 (IX. 5), Formiae, 10 March 49.

[24] *Ad Atticum*, 172, (IX. 6), 1, Formiae, 11 March 49.

[25] *Ad Atticum*, 176 (IX. 9), 2, Formiae, 17 March 49.

[26] *Ad Atticum*, 183 (IX. 15), 1, Formiae, 25 March 49.

[27] *Ad Atticum*, 186 (IX. 17), Formiae, 27 March 49.

[28] *Ad Atticum*, 187, (IX. 18), 3, Formiae, 28 March 49.

"I want to be at Arpinum on the 31st, and then to make a round of my little properties, which I never expect to see again";[29]

"Rome being impossible, I have given my son the white gown at Arpinum as the next best place, to the gratification of my fellow townsmen. Not but what I find everyone both at Arpinum and on the road gloomy and downcast, so sad are the thoughts inspired by this vast mischief";[30]

"... Dionysius arrived at my place bright and early ... You said in the letter I received at Arpinum that he would come ...";[31]

"I think I shall leave here on the 16th, but I may go eitehr to Tusculum or to my house in town; from there perhaps to Arpinum. When I know definitely I shall send you word";[32]

"I must go to Arpinum. My little properties there need my attention ...";[33]

"I know nothing about the house in Arpinum";[34]

"First then I wanted you to know that I am going to Arpinum on 17 May";[35]

"I stayed the night at my place near Sinuessa and am scribbling this letter early the following morning before I leave for Arpinum";[36]

"Now I ask your advice. Do I return to Rome or stay here or should I flee to Arpinum, which offers security?

---

[29] *Ad Atticum*, 188 (VIII.9), 3, Between Formiae and Arpinum, 29 or 30 March 49.

[30] *Ad Atticum*, 189 (IX. 19), 1, Arpinum, 1 or 2 April 49.

[31] *Ad Atticum*, 208 (X. 16), Cumae, 14 May 49.

[32] *Ad Atticum*, 282 (XII.42), 3, Astura, 10 May 45.

[33] *Ad Atticum*, 317 (XIII.9), 2, Tusculum, 17 (?) June 45. Then letters 319-329 follow, written from Arpino or 'in Arpinati', between 22 June 45 and 4 (?) July 45.

[34] *Ad Atticum*, 338 (XIII. 46), 4, Tusculum, 12 August 45.

[35] *Ad Atticum*, 376 (XIV. 22), 1, Puteoli, 14 May 44.

[36] *Ad Atticum*, 378 (XV.1a), 1, Sinuessa, 18 May 44).

I prefer the last, but perhaps Rome is better (?), for fear I may be missed if people come to think that something has been achieved. So how to solve this problem. I was never in a greater quandary";[37]

"I should be grateful for a description of your Shrine of Amalthea, its furnishings and layout ... I have a fancy to make one on my place at Arpinum";[38]

"We shall see you at Arpinum and welcome you in country style since you have scorned our seaside hospitality";[39]

"When I got to Arpinum, my brother came over and we talked first and foremost about you ... Next morning we left Arpinum. I stayed at Aquinum, but we lunched at Arcano (nosti hunc fundum) ...";[40]

---

[37] *Ad Atticum*, 418 (XVI.8), 2, Puteoli, 2 or 3 November 44.

[38] *Ad Atticum*, 16(I.16), 18, Rome, beginning of July 61.

[39] *Ad Atticum*, 36 (II.16), 4, Formiae, 29 April or 1 May 59., at considerable length.

[40] *Ad Atticum*, 94 (v.1), 3, Minturnae, 5 or 6 May, 51. Arcanum is the modern Arce, where Quintus had a villa. This letter adds a colourful picture on the difficult marriage between Quintus and Pomponia, Atticus' sister. Cicero writes: "I come now to the line in the margin at the end of your letter in which you remind me about your sister. This is how the matter stands. When I got to Arpinum, my brother came over and we talked first and foremost about you. From that I passed to what you and I had said between us at Tusculum anent your sister. I have never seen anything more gentle and pacific than my brother's attitude towards her as I found it. Even if he had taken offence for any reason, there was no sign of it. So much of that day. Next morning we left Arpinum. On account of the holiday Quintus had to stay the night at Arcanum. I stayed at Aquinum, but we lunched at Arcanum -you know the farm. When we arrived there Quintus said in the kindest way 'Pomponia, will you ask the women in, and I'll get the boys?' Both what he said and his intention and manner were perfectly pleasant, at least it seemed to me. Pomponia however answered in our hearing 'I am a guest myself here'. That, I imagine, was because Statius had gone ahead of us to see to our luncheon. Quintus said to me 'There! This is the sort of thing I have to put up with every day'. You'll say 'What was there in that, pray?' A good deal. I myself was quite shocked. Her words and manner were so gratuitously rude. I concealed my feelings, painful as they were, and we all took our places at table except the lady. Quintus however had some food sent to her, which she refused. In a word, I felt my brother could not have been more forbearing nor your sister ruder. And I have left out a number

"On the 22nd I received at Arpinum your two letters replying to two of mine ...";[41]

"The people of Arpinum are in an amazing fume about Laterium. Well, there it is! I was very sorry, but 'little he recked my rede'';[42]

"You may hand over the money due to the municipality of Arpinum, the whole of it, if the Aedile L. Fadius claims it";[43]

"Our little Beauty ... accuses me of having been at Baiae-not true, but anyhow ... What business has an Arpinum man with the warm springs? 'Tell that to your counsel' I retorted; he was keen enough to get certain of them that belonged to an Arpinum man' (you know Marius' place of course)";[44]

"As for the dowry, all the more reason for clearing it up ... the Isle of Arpinum allows of a real deification, but I am afraid that its out of the way situation might seem to detract from the dignity of the tribute. So my mind

---

of things that annoyed me at the time more than they did Quintus. I then left for Aquinum, while Quintus stayed behind at Arcanum and came over to see me at Aquinum early the following day. He told me that Pomponia had refused to spend the night with him and that her attitude when she said good-bye was just as I had seen it. Well, you may tell her to her face that in my judgment her manners that day left something to be desired. I have told you about this, perhaps at greater length than was necessary, to show you that lessons and advice are called for from your side as well as from mine".

[41] *Ad Atticum*, 380 (XV.3), Arpinum, 22 May 44.

[42] *Ad Atticum*, 77 (IV.7), 3, Arpinum, ca. 13 April 56. Laterium was one of Quintus' villas, apparently in the modern Fontana Liri, a town contiguous to Arpinum. "Some operation by Q. Cicero on his estate at Laterium, perhaps concerning the diversion of a water course, had annoyed his neighbours" (Loeb Classical Library). Cicero complains that Quintus took very little notice of his advice.

[43] *Ad Atticum*, 393 (XV.15), 1, Astura, 13 (?) June 44.

[44] *Ad Atticum*, 16 (1, 16), 10, Rome, beginning of July 61. The 'little Beauty', or 'pretty boy', is Clodius Pulcher. The 'counsel' is the elder Curio, who acquired Marius' villa near Baiae in Sulla's proscriptions.

is fixed on suburban properties, which however I shall inspect when I get back";[45]

"What an extraordinary coincidence! I got up before daybreak on the 9th, off from Sinuessa, and had reached Tuscan (?) Bridge at Minturnae, where there is a turning on the road to Arpinum, just as it was getting light ...";[46]

"My brother Quintus seems to me to feel towards Pomponia as we wish and is now with her on his estates at Arpinum";[47]

"From Cicero to Brutus greetings.

I have always noticed the particular care you take to inform yourself of all that concerns me, so I do not doubt that you not only know which township I hail from but also know how attentively I look after the interests of my fellow townsmen of Arpinum. All their corporate income, including the means out of which they keep up religious worship and maintain their temples and public places in repair, consists in rents from property in Gaul. We have dispatched the following gentlemen, Roman Knights, as our representatives to inspect the properties, collect sums due from the tenants, and take general cognizance and charge: Q. Fufidius, son of Quintus, M. Faucius, son of Marcus; Q. Mamercius, son of Quintus. May I particularly request of you in virtue of our friendship to give your attention to the matter, and do your best to see that the business of the municipality goes through as smoothly and rapidly as possible with your assistance ... your favour will bind a township which never forgets an obligation. As for myself, I shall be even more beholden than I should otherwise have been, because, while it is my constant habit to look after my fellow townsmen, I have a particular concern

---

[45] *Ad Atticum*, 259 (XII.12), Astura, 16 March 45. Cicero needs the money to settle Terentia's claim, on the occasion of the divorce.
[46] *Ad Atticum*, 423 (XVI.13), 1, Aquinum, 10 November 44.
[47] *Ad Atticum*, 2 (1.6), 2, Rome, shortly after 23 November 68.

and responsibility towards them this year. It was my wish that my son, my nephew, and a very close friend of mine, M. Caesius, should be appointed Aediles this year to set the affairs of the municipality in order -in our town it is the custom to elect magistrates with that title and no other. You will have done honour to them, and above all to me, if the corporate property of the municipality is well managed thanks to your good will and attention. May I again ask you this favour most earnestly"?;[48]

"When our friend Lupus arrived from you and spent some days in Rome, I was where I thought I should be safest";[49]

"In a separate letter I have recommended the representatives of the people of Arpinum jointly to your favour as warmly as I was able. In this letter I am recommending Q. Fufidius, with whom I have all manner of friendly ties ... He ... served with me as Military Tribune in Cilicia";[50]

"I had just got in from Arpinum when I was handed a letter from you";[51]

"... if you agree, please use the country houses which will be the farthest away from army units. The farm at Arpinum with the servants we have in town will be a good place for you if food prices go up";[52]

---

[48] *Letters to Friends*, Cicero to Brutus, letter 278 (XIII, 11), 1-3, Rome, 46. The city of Arpinum, after the battle of the Campi Raudii, near Vercelli, had been donated by Marius some territories in Cisalpine Gaul from which it was still receiving revenues at the time of Cicero.

[49] *Letters to Friends*, Cicero to D. Brutus, 353 (XI.5), Rome, 9 December (or shortly after) 44. Cicero refers to Arpinum, the place where he 'should be safest'.

[50] *Letters to Friends*, Cicero to Brutus, 279 (XIII.12), 1, Rome, 46.

[51] *Letters to Friends*, Cicero to M. Fabius Gallus, 209 (VII.23), 1, Rome, beginning of December (true calendar) 46.

[52] *Letters to Friends*, Cicero to Terentia, 155 (XIV.7), aboard ship, Caieta Harbour, 7 June 49. Cicero was leaving Italy to join Pompey's army at Dyrrachium.

"Tomorrow I intend to stay at Laterium, and then, after five days in the Arpinum area, go on to Pompei";[53]

"Escaped from the great heat wave (I don't remember a greater), I have refreshed myself on the banks of our delightful and salubrious river at Arpinum during the Games period ... I was at Arcanum on 10 September. There I saw ... the stream which they are bringing over not far from the house. It was flowing merrily enough, particularly in view of the severe drought ...";[54]

"The boys are well, keen at their lessons and conscientiously taught ... The finishing touch to both our houses are in train, but your country residences at Arcanum and Laterium are already but complete ...".[55]

b) Let us transcribe the passages where Cicero mentions Sora.

"Q. Valerius Soranus, "the most erudite littérateur of all who have the Roman citizenship ...";[56]

"Appius ... your colleagues used to laugh at him and call him at one time 'a Pisidian' and at another 'a Soran';[57]

"Quintus and Decimus Valerius of Sora, neighbours and friends of mine".[58]

We can now draw some conclusions from the above citations, making it clear, first of all, that, like Sir Isaac Newton, 'hypotheses non fingo', I do not feign hypotheses.

1) As we see, Cicero mentions numberless times Arpinum, but Sora just on four occasions, in one of these occasions with a mocking remark. And yet the Convent of San Domenico, where, supposedly, his house was, is at half a mile from Sora.

This seems to suggest that he had very little to do with Sora.

---

[53] *Letters to Quintus*, 10 (ii,.7), Rome, shortly after 15 May 56.

[54] *Letters to Quintus*, 21 (III.1), Arpinum and Rome, September 54.

[55] *Letters to Quintus*, 23 (III.3), 1, Rome, 21 October 54.

[56] *De Oratore*, III, xii, 43.

[57] *De Divinatione*, I, xlvii, 105. In note the translator, W. A. Falconer, adds the following words: "the Sorans, who lived in Sora, a small town in Latium, were noted for their superstition".

[58] Brutus, XLVI, 169. See furthermore text corresp. with no. 10.

Where he calls Quintus and Decimus Valerius of Sora his 'neighbours', this, by contrast, suggests vicinity.

From the text of the *De Legibus* one should conclude that the house was where the convent now stands. This was Card. Baronius' conclusion, as well as Francesco D'Ovidio's, who entertained no doubt at all on this hypothesis.[59] Nevertheless, no doubt Cicero's pages are in large part fictional, in the sense that the author in the opening section of book II aims at emphasizing the 'Quercia Mariana' and his own poem, the *Marius*. Endless discussions concerning the point of departure of the walk of the three friends have led to no plausible conclusion.

The Roman stones that can be found in the area around the Convent are all gravestones. This demonstrates that in the same place there was a cemeterial area, not a Roman Villa. Excavations have brought to light stones and pieces of sculpture from that burial ground, which was built, along the *via publica*, in the I and II centuries A.D.

Many of these stones have been reused, and some pieces of sculpture have been bricked up on the walls of the Convent.[60]

Curiously enough, there are no stones at all, both in San Domenico and in Arpino, that could make reference to the Tullii, i.e., to Cicero's family.[61]

One objection to this view is that Cicero's house could have been submerged by the floods of the two rivers, the Fibrenus and the Liris, and the Church and the Convent could have been built above it. Nevertheless excavations that took place in the past have given no evidence for this. The archaeologist Alessandra Tanzilli who, years ago, was able to enter the sepulchral chamber, and the geologist Marilena Rufo, assure me that the level of interment of the area has been very modest, and that the church[62] was built in part on made ground. The church was built in the eleventh century A.D., in the years 1011-1130, and during the last ten centuries the variations of the two rivers have been minimal. This is proved by

---

[59] See F. D'Ovidio, 'Di dove era l'Arpinate?', in 'Varietà Filologiche: scritti di filologia classica e di lingua italiana', in *Opere Complete*, X, pp. 117-151.

[60] See photo no. 11.

[61] Gian Luca Gregori, prof. of 'Roman Antiquities' at the University of Rome 'La Sapienza', makes this point.

[62] See photo no. 10.

the Roman remains, for example the Ponte Marmore, or Marmone, which crossed the Liris,[63] and of which just one arch is still extant. There are furthermore calcareous banks that, by their nature, could not have been eroded in two thousand years. Last, floods have certainly taken place, but not such as to endanger the church, and in the previous ten centuries there must have been the same modest floods and minimal variations, not enough to change substantially the landscape.

The easiest conclusion, that one can draw from Cicero's words, according to whom "the island ... cuts the Fibrenus like the beak of a ship, and the stream ... immediately (*statim*) plunges into the Liris", is that the house was where the convent stands now. This conclusion has been commonly accepted for centuries but, as we have said, it cannot be proved.

If one accepts it at face value, one should say that Cicero's island is the one at 100 yards from the Convent.[64] This 'island' is no longer an island today, -and maybe it was not in the past- because the Fibrenus does not 'come quickly together again', but plunges into the Liris divided into two separate branches, one of them, the northern one, being clearly artificial. One can well suppose that twenty centuries ago the two branches used to 'come quickly together', but this supposition doesn't help, because, as said above, and in any case, there is no archaeological or epigraphic evidence that the house was exactly on the spot where the convent now stands.

2) No doubt, in twenty centuries the morphological configuration of the area has undergone changes. The Fibrenus is a river constant in its flow of waters, and is only eight kilometres long. The Liris, by contrast, is a *torrential* river, in the sense that its waters contract in summer but flow in large quantities in winter, with heavy rains, causing floods and erosion of the banks. Therefore, in twenty centuries it has certainly changed its course, at least in part, if not in the principal flood, probably concerning some minor branches. In a similar way the river Rapido, near Cassino, in South Latium, now flows in a bed distant half a mile from the bridge under which it used to flow twenty centuries ago. A moderate knowledge of geology makes it clear that rivers, at least in part, can change their course.

---

[63] See photo no. 16.
[64] See photo no. 8.

A study to be mentioned on these problems is the one by Modesto Galante, 'A Carnello l'Amaltheum e la villa natale di Cicerone', along with 'L'ultimo studio sulla casa paterna di Cicerone'.[65] Both these studies are the result of accurate explorations in the area, but the conclusions are not supported by sound geological studies, and on the whole they are not very convincing. According to the author, Cicero's house was at Carnello, at one mile from San Domenico, just behind the river and the Church, but this conclusion, although intriguing, may perplex the reader. There is, however, at Carnello, another *insula*, or island,[66] that could be the one mentioned by Cicero, but the conclusions that Galante draws appear rather risky and, until now, there is no archaeological or epigraphic evidence for them.

3) The villa apparently was not at San Domenico but somewhere else, in the *ager*, or in a piece of land, nearer to the city of Arpinum. And Cicero's words suggest vicinity, even intimacy with the city, where his family could have had a house: although there is no record for this.

A Roman Villa[67] *recently came to light in Via Sant'Altissimo, contrada Pagliare at Carnello di Arpino. According to the archaeological authorities it was built in two instalments, and was in use from the first century B.C. to the mid first century A.D.*[68] Also in this case there was a modest level of interment, because the floods of the Fibrenus, which no doubt took place during the centuries, were not such as to bury both the ground and the building at great depth. That is the reason why the conservation of the remains has been rather modest.

All this corresponds to the description that Cicero gives of his natal house, that was built in two instalments: "*Yonder you see our homestead as it is now -rebuilt and extended by my father's care ... it was on this very spot ... that I was born ... when the homestead, according to the old custom, was small ...*".[69]

The 'other' island of the river Fibrenus is quite close, at two hundred metres from the house. It remains to translate adequately

[65] See Bibliography.
[66] See photo no. 9.
[67] See photo no. 15.
[68] See the journal 'Il Ponte', Officina della Cultura, Ottobre 2014, pp. 42-3.
[69] See above, text corr. with no. 6.

the Latin adverb *statim*, of the Fibrenus that *statim precipitat in Lirim*. According to the *Oxford Latin Dictionary* 'statim' means 'immediately', 'at once', but also 'constantly', 'regularly', and what did Cicero actually mean?

Last. The river Fibrenus until the end of the Republican Age represented the northern boundary of the *ager*, or countryside, of Arpinum. Therefore, the villa was in the boundaries of Arpinum, not in those of Sora, and Cicero was an Arpinas, not a Soranus.

To be sure of the identification, one or more inscriptions should be traced, because in the same area, at least in theory, there could have been more villas of local magistrates, but until now no inscriptions and no more villas have come to light. In the last period of the Republic there were apparently less than six million inhabitants in Italy, large spaces were therefore available, and Cicero's villa had a large property around it. This is an argument in favour of the identification.

The area has been placed under preventive attachment by the archaeological authorities, and further excavations, that hopefully will take place in a near future, could confirm the find.

Let us now consider, for a moment, the Arpinum of our infancy and early youth. There was then, if not families, if no longer descendants, certainly the memory of families who, although living in the city, owned large estates in the low part of the land, on the plain. And everything suggests that Cicero's family was one of this stock of families, who had moved to the plain because, owing to their business as *fullones*,[70] or fullers, they needed the use of water: of the water of the river Fibrenus, in particular.

We have transcribed, here above, the passages which suggest Cicero's contiguity with the city of Arpinum and, apparently, a real belonging to it.

To these passages more passages could be added, where he mentions the city indirectly.

---

[70]This word, indicating a family business, was used with a derogatory sense against Cicero by arrogant members of the Roman nobility. Arpinum was in any case, until two centuries ago, a centre for the manufacture of wool, owing to the breeding of sheep, which was a profit making activity thanks to the wide extent of the land, especially on the mountainous part of it. Cicero's family was apparently in this kind of business, and the orator's immediate ancestors could have in part moved to the plain.

For example, on some occasions he mentions the Amaltheum, that he had in mind to build in Arpinum: "My Amalthea awaits you and needs you";[71] "I shall not neglect your reminder about Amalthea";[72] "If I am let live in peace, I shall not drag you away from Amalthea".[73]

Furthermore, and this seems of relevant importance, he dates his letters from Arpinum sometimes as 'Arpini', which means from the city and, alternately, 'in Arpinati', which refers to the country, where he owned the villa. From the date of his letters we can infer the periods that he spent in his property in Arpinum, or in the city.

We have emphasized, transcribing them in bold, in particular the phrases which suggest close ties of the orator with the city of Arpinum. We must transcribe them again, here.

So there is Atticus saying that he had

> "*the idea that there was nothing in this vicinity except rocks and mountains*, and both your speeches and poems encouraged me in that opinion";[74]

> "*all our rugged countryside (aspera et montuosa)*";[75]

> "*my native hills, the cradle of my birth*";[76]

> " ... Lucius Antonius, whose cruelty I escaped only *under the protection of the walls and gates of my native town*";[77] i.e., the city of Arpinum: far from San Domenico and far from Sora;

> "*I wanted to give Marcus his white gown in Arpinum*";[78]

> "*I have given my son the white gown at Arpinum as the next best place, to the gratification of my fellow townsmen*";[79]

> "*should I flee to Arpinum, which offers security??*";[80]

---

[71] *Ad Atticum*, 21 (ii, 11), Antium (?), ca. June 3 (?), 60.

[72] *Ad Atticum*, 27 (II, 5), Antium, early April 59.

[73] *Ad Atticum*, 40 (II.20), 2, Rome, ca. mid-July 59.

[74] *De Legibus*, II, i, 1, II, iii, 6-7.

[75] *Pro Plancio*, VIII-IX, 20-22

[76] *Ad Atticum*, 35 (II, 16), Formiae, ca. 28 April 59.

[77] *Phil.*, 12.20.

[78] *Ad Atticum*, 172, (IX. 6), 1, Formiae, 11 March 49.

[79] *Ad Atticum*, 189 (IX. 19), 1, Arpinum, 1 or 2 April 49.

[80] *Ad Atticum*, 418 (XVI.8), 2, Puteoli, 2 or 3 November 44.

"... while it *is my constant habit to look after my fellow townsmen*, I have a particular concern and responsibility towards them this year. It was my wish that my son, my nephew, and a very close friend of mine, M. Caesius, should be appointed *Aediles this year* to set the affairs of the *municipality in order*; I have recommended the representatives of the people of Arpinum jointly to your favour as warmly as I was able".[81]

While he is leaving Italy to join Pompey's army, while on board ship, at Caieta, he recommends Terentia and Tulliola to use "the country houses which will be the farthest away from army units. The farm at Arpinum *with the servants we have in town*[82] will be a good place for you...".[83]

"I was where I thought I should be safest".[84]

This is the literary evidence available, and needs to be emphasized. As far as we know, until now this had not been sufficiently done.

This helps to sum up the present state of things, concerning the relation of Cicero with the city of Arpinum and his family's homestead in the country. Apparently, the homestead was not far from the town, but the place where it actually was, is still an open question. Cicero's close relation with Arpino-city should, however, no longer be in question.

---

[81] *Letters to Friends*, Cicero to Brutus, letter 278 (XIII, 11), 1-3, Rome, 46. The city of Arpinum, after the battle of the Campi Raudii, near Vercelli, had been donated by Marius some territories in Cisalpine Gaul from which it was still receiving revenues at the time of Cicero.

[82] In the town of Arpinum.

[83] *Letters to Friends*, Cicero to Terentia, 155 (XIV.7), aboard ship, Caieta Harbour, 7 June 49. Cicero was leaving Italy to join Pompey's army at Dyrrachium.

[84] *Letters to Friends*, Cicero to D. Brutus, 353, 1, (XI.5), Rome, 9 December (or shortly after) 44. Cicero refers to Arpinum, where he 'should be safest'.

Bibliography

(c/o Alessandra Tanzilli)

Alberti, L., *Descrittione di tutta l'Italia* (Venezia, 1557).

Aurigemma, S., *Configurazione stradale della regione sorana nell'epoca romana*, in *Per Cesare Baronio – Scritti vari nel terzo centenario della sua morte* (Roma, 1911), pp. 493-547.

Baronio, C., *Martyrologium Romanum* (Venetiis 1609).

Baronio, C., *Annales Ecclesiastici*, XVI (Lucae, 1744).

Branca, C., *Memorie storiche della città di Sora* (Napoli, 1847[1], Bologna, 1982[2]).

Cassoni, M., *Sguardo storico sull'Abbazia di San Domenico in Sora* (Sora, 1910).

Cassoni, M. *La villa natale di Marco Tullio Cicerone e i vari possessori della medesima* (Sora, 1911).

Cerqua, M., Cerrone, F., Pantano, W., *La necropoli imperiale di S. Domenico a Sora (Frosinone)*, in *Atti del Convegno, Settimo Incontro di Studi sul Lazio e la Sabina* (Roma, 9-11 marzo 2010), a cura di G. Ghini, pp. 445-452.

Clavelli, B., *L'antica Arpino* (Napoli, 1623).

Coarelli, F., *Insula Arpinas: il sito della casa di Cicerone*, Bollettino dell'Unione di Storia ed Arte, 4, anno X n. s., 1967, pp. 77-79.

De Nino, A., *Sora*, in *Notizie degli Scavi d'Antichità* (1879), pp. 117-119.

D'Ovidio F., *Di dov'era l'Arpinate*, Atene e Roma, II, 1899, 11, pp. 200-218.

Fortini, P., *Charles Kelsall in visita al luogo di nascita dell'illustre oratore*, in *Paese mio* (settembre 2010), p. 8.

Galante, M., *A Carnello l'Amaltheum e la villa natale di Cicerone*, Ciceroniana, I (Firenze 1959), pp. 3-18.

GALANTE, M., *L'ultimo studio sull'ubicazione della casa paterna di Cicerone* (1999).

Giannetti, A., *La casa natale di Cicerone al centro di una polemica senza tempo, Rassegna del Lazio* (12, 1974), pp. 55-59.

Hoare, C., *A Classical tour through Italy and Sicily* (London, 1819).

Holstenius, L., *Annotationes geographicae in Italiam antiquam P. Cluverii* (Roma, 1666).

Ippoliti, L., *Il luogo di nascita di Marco Tullio Cicerone* (Arpino, 1936).

Kelsall, Ch., *Classical Excursions from Rome to Arpino* (Geneva, 1820[1]).

Lisi, G., *Historia Sorana* (Roma, 1728).

Pierleoni, G., *Il paesaggio del De legibus*, Bollettino del museo civico di Arpino , I, 1, (Arpino, 1912).

Pisa, Maria G., *Arpino nelle opere e nell'animo di Cicerone, tesi di laurea, Istituto Universitario di Magistero 'Suor Orsola Benincasa'*, Napoli, Anno accademico 1941-42.

Pistilli, F., *Descrizione storico-filologica delle antiche e moderne città e castelli esistenti accosto i fiumi Liri e Fibreno* (Napoli, 1798).

Rizzello, M., *Carnello e la via del Fibreno* (Casamari, 1990).

Romanelli, D., *Antica topografia istorica del regno di Napoli* (Napoli, 1819), vol. III.

Schmidt, O. E., *Cicero's Villen*, in *Neue Jahrbücher für das Klassische Altertum*, (I, 1899).

Schmidt, O. E., *Arpinum* (Arpino, 1907).

Tanzilli, A., *Antica topografia di Sora e del suo territorio* (Isola del Liri, 1982).

Tanzilli, A., Grimaldi M., *Museo della media valle del Liri – Sora, guida alla sezione archeologica di età romana* (Isola del Liri, 2009.

Tanzilli, A., *Materiali funerari editi e inediti a Sora, Vicalvi e Casalattico, Quaderni Coldragonesi*, 4, a cura di A. Nicosia (Colfelice 2013), pp. 17-25.

Photographic Illustrations

1) The city of Arpinum, panorama (foto Folchetti, Arpino)

2) The Liceo Classico Tullianum (foto Folchetti, Arpino)

3) Arpinum, Cicero's "walls and gates of my native town"
(foto Folchetti, Arpino)

4) The rapids of the river Fibrenus, at two hundred metres from
San Domenico Abbey (foto Folchetti, Arpino)

5) The cloyster of San Domenico Church, with both Roman and
Medieval columns (foto Folchetti, Arpino)

6) A spiral column in San Domenico Church's cloister, apparently
a milestone of Emperor Massentius' (278-312) age
(foto Folchetti, Arpino)

7) The principal branch of the Fibrenus empties into the Liris
(foto Folchetti, Arpino)

8) The Fibrenus is cut 'like the beak of a ship' at two hundred metres from San Domenico Abbey (foto Folchetti, Arpino)

9) The beak cutting the Fibrenus and forming the 'other' island at
Carnello (foto Carlo Scappaticci, Arpino)

10) San Domenico Abbey, façade (foto Folchetti, Arpino)

11) Doric funerary sculpture, south side of the Church
(foto Folchetti, Arpino)

12) San Domenico's apse resting on frameworks of sepulchral
monuments with a circular lay-out, 1st-2nd centuries AD
(foto Folchetti, Arpino)

13) Stones from Roman sepulchral monuments, in the North
Garden of the Church (foto Folchetti, Arpino)

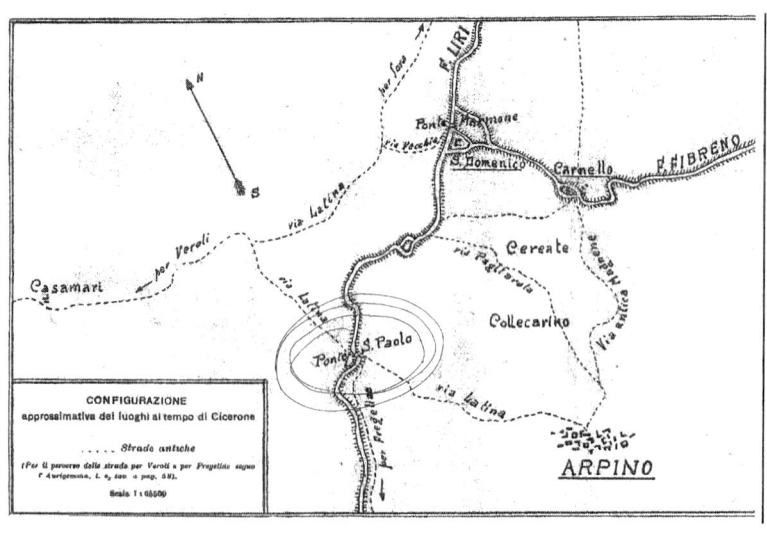

14) Map of the places, after Aurigemma

15) Roman Villa that recently came to light in contrada Pagliare,
Via di Sant'Altissimo, at Carnello di Arpino.

16) Ponte Marmore (foto Mauro Tomaselli, Carnello)

Index

Vincenzo Merolle, of the university of Rome "La Sapienza", Faculty of Political Sciences, has edited, for the Pickering Masters, Ferguson, *Correspondence*, 2 vols, (London 1995), *Ferguson, Manuscript Essays* (London, 2006) and, with Eugene Heath, of New York State University at New Paltz, *Adam Ferguson: History, Progress and Human Nature* (London 2007), and *Adam Ferguson: Philosophy, Politics and Society* (London 2008).

He is the editor of *2000. The European Journal*. He is furthermore the editor of the *Directory of Scholars in European Studies* (see on the internet under www.directoryeuropeanstudies.com) and the author of *The European Dictionary*, vol. 1, (LOGOS Verlag, Berlin, 2013).

Among his books in Italian deserving mention are: *Gramsci e la filosofia della prassi* (Bulzoni, Rome, 1974); *Adam Smith: Politica e Società* (Bizzarri, Rome, 1984); *John Millar, Letters of Crito e Letters of Sidney*, text in English with introduction and notes in Italian (Giuffré, Milano, 1984); *Saggio su Ferguson. Con un Saggio su Millar* (Gangemi, Rome, 1994).